Living wi

Twenty-five Years of Life on the Mountain

Diana Coogle

Laughing Dog Press
Blue Ridge, Georgia

These essays were first broadcast
on Jefferson Public Radio, Ashland, Oregon.

❖

Sincerest appreciation

to Rosemarie Perrin and Amy Belkin,
for their critique of the manuscript
and especially to Susan Antinori
for her careful and astute suggestions,

to Lucy Edwards at Jefferson Public Radio
for her friendship, her support,
and her incomparable skills in producing my commentaries,

and to Ryanne Webb for computer assistance.

❖

Laughing Dog Press
Blue Ridge, Georgia 30513
*
P.O. Box 3314
Applegate, OR 97530
dcoogle@rogue.cc.or.us

❖

This book was set in Times, 10 point.

To my sister Sharon,

who has given generously of her time,
artistic expertise, and supportive affection,
not only to me and not only for this project,
but to many people on many occasions.

Contents

Introduction

I started building a little house in the mountains in 1974, when my son, Ela, was about two and a half. His dad (Dan) and I had just separated. I don't know what possessed me to think I could build a house except that I needed one and the land I was living on didn't have one. I had the confidence of having learned something about carpentry by hanging out with my father in his basement shop, and I think I may have read a book, too. I had a hammer and a saw; what more did I need? Anyway, I was only going to build a very small house, 10 feet by 12 feet with a 12-foot ceiling, since a friend had told me that a high ceiling would keep a small space from being oppressive in the winter. The south side of the peak roof would be skylights all the way across.

Ela and I lived in this little tiny house in the mountains without a car and without a telephone and without electricity, which meant without a television and without a stereo player and without a refrigerator or any fancy kitchen equipment and without electric lights and it goes without saying without a computer and computer games. But it wasn't what we didn't have that defined our lives but what we did have - an endless forest to play in, wild animals and birds at our doorstep, a moon that shone full through the skylights, and time that was our own. Though we lived far from town, we had friends nearby who lived as we did, and Ela lived with his dad, just eight miles on the other side of the mountain, every other month while he was very young and then every other year.

It was a life that enhanced the senses. I was aware of the thrum of insects by day and the warbling of raccoons and whistling of deer in the evenings. In June I often caught sight of the yellow and red flash of a western tanager in my cherry tree; on clear, moonless nights I gazed at jillions of stars in my patch of sky over the mountain, and all summer long I could watch the button-sized knobs on my apple tree grow into large, firm, ripe apples. The smell of green fir on a walk alerted me to a tree recently felled by a storm, and dinner always tasted better after such a walk in fresh mountain air.

After eight or ten years, I added on to the house, this time with a carpenter to help me. I bought a small generator to run his power tools, and, several years later, I also got a telephone (the man installing the cable said, "Lady, I've been to some out-of-the-way places, but this takes the cake") and a hot water heater for an outdoor shower and an electric sewing machine and a laptop computer to replace my old typewriter. I also bought a car, though I never did put in a driveway. That's about it. Ela has grown up and been to college and moved to Seattle and gotten married and developed his own career as a musician/sculptor/performer. I still use kerosene lamps and mix cake batter by hand; I still park the car down the hill and walk up through the forest to get home, and I still play my guitar if I want to listen to music other than the hum of the wind or the songs of birds. I still define my life in terms of what I have rather than what I don't have, and what I have is pretty much the same as it was twenty-five years ago.

One time while I was living in the one-room house, an acquaintance who was interested in the way I lived came to visit. He stayed for dinner and on into the summer evening.

"You should write about all this," he said, sweeping his arm to indicate not only my little house but the forest outside the window and the garden I had planted and all the wilds I lived among.

"Well, I have," I said, and I explained that I had recently read a book of short essays by a French writer named Alain who had taken as an exercise to write an essay a day, without revisions. Inspired, I had taken on the same exercise, using my house and my life here as subject matter - except I couldn't adhere to the rules because I like revising, and so I was always going back to the previous day's essay to work on it.

But it was just a writing exercise. Who would be interested?

"Lots of people," he said.

Buoyed by that kind of encouragement, I revised and polished and revised and polished some essays that I called "Nine Rules for One-room Cabin Living" and submitted them to the local public radio station. Tom Olbrich, at Jefferson Public Radio, said, well, yes, maybe they would try this. Later he told me that radio people are leery of writers who want to read their own works on the air because not every good writer is a good reader, but that he liked my writing and I had an interesting voice for radio, with its hint of the South.

I read my commentaries weekly for seven years, at which point, thinking I was beginning to sound like a parody of my own style, I quit to work on longer essays and to write plays. I missed my radio work, though, and in 1996, refreshed by my absence from this type of writing, I reapplied to the station and to its news director, Lucy Edwards. Ever since then, I have again been reading my commentaries weekly. Most of the essays in this book come from this later time period, though a few snuck in from my first seven years on the radio.

When I started putting together this collection, I had in mind a theme vaguely centered around the five senses of sight, hearing, taste, smell, and touch, which have been so much a part of my life, but as I reread my essays and organized and reorganized them, I realized that five other senses are just as important to me. Living on the mountain has given me a quintessential sense of place, of home. It has given me a sense of nature that has become part of my flesh and blood, and it has set my imagination flying with a strong sense of fun. This contemplative life on the mountain has deepened my sense of other and increased my love and compassion for family, friends, and humanity while at the same time deepening in a spiritual way my sense of self.

The table of contents refers to this latter meaning of the five senses, and that's how this book is organized, but as you read, I hope you'll see the other five senses, too, imbuing this work with their vitality. After all, it takes all ten senses to give life its precious glow.

x

Prologue

Advice

Written by my son Ela, when he was 24, and put into my stocking on Christmas Day, 1996.

Live on a mountain; build your own house.
Have a proliferation of creative ideas.
Spend much time exploring words and spinning many tales.
Keep cats.
Make many flowers bloom.
Make many magic meals for one or many.
Walk alone in the forest.
Have many wonderful friends.
Frequently finger a fine guitar.
Be like good wine - get better with age.
Have a distinctive laugh to be remembered for good humor.
Swim silently at sunset; swim swiftly at sunrise.
Spin like the spider many garments of gold.
Be a teacher of intelligence.
Inspire.

A Sense of Home

Why I Live in Oregon

I served on a panel of writers once who were asked to answer the question, "Why do you live and work in Oregon?" The question assumes I have a choice about the matter. I'm not sure I do.

The answer begins in 1969 when I was living in Pikeville, Kentucky, and was newly in love. My boy friend, Dan, and I joined the sweep west of that era. When we got to Oregon we said we were looking for a place in the country close to a cultural center, and people said, "Oh, you want to go to Ashland," and so we went to Ashland and told people we wanted a place in the country, and they said, "You want to go to the Applegate," and so we went to the Applegate and found a barn we could turn into living quarters, and then a couple of years later we weren't together any more and I moved onto a piece of land two miles up the mountain from that barn and started building my house.

I knew even before I started building the house, though — even just shortly after I got to Oregon - that I had come home, and that was not only because I am not a gypsy soul and was ready to have a home somewhere, anywhere, but because something about Oregon — its mountains and rivers, yes, but something more mystical than that — told me I was home. When I did start building my house, my friend Howdy told me how to set the main poles in the ground, but he didn't mention any protective covering, so I said, tentatively, "But, Howdy, shouldn't we creosote the poles?" and he said, "Oh, they'll last five years, and no telling where you'll be in five years." But I had a feeling I would be in that same house five years from then and twenty-five years from then, which is now, and thirty-five years and forty. I had a feeling I wasn't going to be leaving.

As I built my house, my roots sank into the earth. They grew fast and they dug deep. They wrapped around rocks and clung to the hillside and tangled around the roots of trees. At the same time, the landscape was entwining itself around me like the wild grapevine I

3

transplanted onto my land twenty-five years ago. Last year I replanted it in front of my pantry, which has a shed roof that meets the house just under my bedroom loft, where the skylight window opens upward in the summer.

By now the grapevine has spread across the entire front wall of the pantry and up onto the roof. Two long arms are moving jauntily towards the open skylight. By tonight or tomorrow night they'll be crawling over the bed, investigating the inside of the house, and when I go to sleep at night, they'll lay their arms over my neck, entwine their tendrils between my fingers and toes, and serpentine their way over my legs and thighs.

So I live and work in Oregon not for reasons the mind gives me but because the landscape has entwined itself around me, even as my roots have dug into the Oregon soil.

The View out the Window

My writing desk sits in a small bay with three windows: the west one at my back, the south one to my right, and the east window, which opens onto the deck, just in front of me. This is the window I look out of most often. Through it, I see my honeysuckle vine, a gift from my friend Maren, pouring over the deck railing just beyond the window. Soon the dusk-loving, deep-throated, humming-bird-attracting, bee-suckling blossoms will be aggressively throwing a honey-baked scent through the open window, sweet and redolent of romance. Then I will see needle-beaked hummingbirds with green-spangled breasts zooming in for a drink.

On the deck itself just outside my window I see pots of flowers: pink, scarlet, and fuchsia anemones; purple and yellow pansies; a fuzzy blue ageratum; the greenery of petunias too young to blossom; and what will be trailing cascades of lobelia in this, my deer-defiance garden.

In mid-air outside my window I see the long, silver-grey tubes of the wind chimes Ela made for me, dangling from the unseen upper deck, waiting, like a piano, for the fingers of the breeze to make its music.

Just behind the deck I see the close-cropped limbs of the apple tree. In a few weeks it will be in full bloom, each blossom blushed with pink, like the cheeks of a bride. In a later season, I might leave my writing to walk onto the deck and pick an apple, which I'll eat leaning over the railing and looking at the garden below or at Humpy Mountain in the close distance.

Humpy is the dominant scene from my window. Season by season, I watch it: the snow-lace of early winter, the thick white hat of heavy snow, the golden glow of sunset, or, as now, the blue wash-out in the backlighting of the sun.

I also see, out this window, my plum tree, giant showcase of the spring with its ocean-foam froth of pale pink blossoms, so big it spills from the eastern window in front of me to the southern window on my right and so potently perfumed I can smell it not only from my desk, now, through the open window, but even from down the path, a five-minute walk away, when I step out of my car. Bees, attracted by the increasingly intensified scent as the sun warms the blossoms, have begun their work. Later, in the afternoon sun, each bee will be set aglow against the dark background of the forest, the entire multitude whirling around the plum tree like a chaotic flurry of snowflakes. Their day-long hum is a sitar drone under the melodies of wind chimes and hummingbirds.

I see wasps fly in and out the window, and just now I am startled by a crashing in the woods below me. I look up to see two deer flash through the trees, then stop, almost invisible in their statuesque stillness. Then, tip-toeing in jerks, their ears twitching, they step into the clearing below my house. They stalk under the plum tree and up the garden path under the window to my right. As I turn to look through the window behind me, I see them walking across my lawn, past my house, and on into the woods beyond.

How to Lay a Propane Pipeline

Without a driveway to my house, whatever gets here has to be carried up the hill by hand. Various people at various times have suggested various ways to end this labor: hot-air balloons, donkeys, winches, pulleys and cables, but I have never bothered with any of those things. I really haven't minded carrying things up the hill – except for the five-gallon propane tanks for my stove and water heater. They're heavy and awkward, and year by year they've gotten heavier and the trail steeper and my temper shorter. Finally, an engineer friend had a brilliant idea: that I put in a pipeline from where I park my car – a 200-foot pipeline not along my curvaceous path but straight up the hill to the house.

And so I did, and it has been such a success I'd like to help other people who might be in a similar situation. If you, too, want to end the decades'-long labor of carrying propane cylinders up the hill to your house, here's how:

Have good friends. This is the key to the whole project.

Be grateful to the friend who suggested the idea because that was the beginning of the end of the labor.

Ask Suburban Propane in Grants Pass for help. They'll enter enthusiastically into the project, saying you deserve this after twenty-five years. Be grateful for their free consultation at your remote mountain house. Be grateful that the store's manager, noting your gasp at the final bill for pipeline and fittings, finds ways to reduce the cost. Tell them they are the best folks in town to work with.

Organize a work party for Saturday, calling on those good friends I mentioned above. When one especially good-hearted one calls you up and says he would like to dig the ditch on Wednesday before the work party, protest: "But, Tom, digging the ditch is the hard part," and when he says, "That's why I want to get it done ahead of time," protest again: "But that's why I asked others to be here, too," and when he says, "It'll be easier on Saturday if the ditch is already dug," protest again: "But, Tom, I won't even be here to help," and when he says, "That's all right," accept and be grateful.

When this same Tom, who dug half the ditch by himself on Wednesday, comes up on Friday to finish the job, stay with him as he flings pick and shovel over roots and stones. Say you'd like to help. When he declines, say, after a while, you think you'll go work in the garden for a bit. He will quickly put a shovel in your hand and suggest you scoop dirt out of the ditch as he digs. Bend your back to the work as he has his, and because he has been working all night at the mill and then all day here with you, DON'T YOU DARE COMPLAIN THAT YOUR BACK IS HURTING! Just keep working. Feed him asparagus penne salad with gorgonzola cheese for lunch.

When your friend Joe comes up the next day to connect the pipeline, see him as a magician with fittings and wrenches. Apologize that big tall he has to bend and scrunch to work in your made-for-your-size, little house. Feed him asparagus tacos for lunch.

Fill in the ditch well, breaking up clods and closing air pockets.

Turn on the gas. Cook dinner. Go to bed.

When, the next morning, you discover the tank is out of gas, avoid the overwhelming guilt that you must have nicked the plastic pipe while you were filling in the ditch. Don't spend the day dreading telling Tom the whole thing will have to be dug all over again. He'll simply say, "I don't think you could have cut the pipe. I'll come up tomorrow, and we'll find the leak."

With Tom, search for the leak. Spray soapy water on all the joints; put plumber's tape on the two that fizzed, and then, when you turn the tank on and discover the gas is still leaking, climb despondently up the hill again. When you get back to the house, Tom will say he thinks he smells gas. Watch, together, as a worm comes clawing and gasping out of the ground. Cry, "It's the canary in the coal mine!" and carry it to the garden as Tom unearths the end of the pipe. Gas will come spewing out; Tom will fix the leak. Be immensely grateful. Fill in the ditch yourself. Give him homemade quiche and apricot bar cookies for lunch.

Take a hot shower with the ultimate satisfaction of knowing you will never have to carry a propane tank up the hill again.

Know that you have the best friends in the world. And be grateful.

Building a Woodpile - Theoretically

Every year it is the major task of autumn to haul to the house two cords of firewood from where they are dumped at the top of the hill on a level with and about 100 yards from the house. Wheelbarrow load by wheelbarrow load, the chaos of a pile of firewood thrown off the truck must be recreated as a neat stack of firewood next to the house. Theoretically neat. It is a task heavily dependent on theory.

Theoretically I get the job done before the rains start. In actuality, I am often pushing that wheelbarrow in the rain.

Theoretically, the woodpile at the house is neatly squared and measures a precise 8x4x2, cord by cord. In actuality, it is impossible to build square ends. The wood has to lean in to hold itself up, so what should be a cubic figure turns out pyramidal with a flattened top.

Theoretically, the top of the woodpile is level and even; in actuality, some top piece sticks up just a bit too high and demands another piece to even things out - which sticks up a little too high, demanding another piece, and so on. It's my father's biscuit-and-gravy dilemma: he's got gravy left, so he needs another biscuit to sop up the gravy, but then he has some biscuit left and has to take more gravy, but having misjudged the amount of gravy, he takes another biscuit, until he ends up leaving gravy on his plate, just as I accept the less than perfect firewood stack.

Theoretically because the pieces are interlocked like puzzle pieces, not stacked like a tower, the woodpile rises as plumb and tight as a wall. But when I look at it broadside, I see it caves inward like a kid who has learned to suck his stomach up under his ribs. A firewood stack with a concave center is sure to break into a hundred pieces, crumpling order to chaos, and, worse, much of the chaos could tumble down the hill on the edge of which I've built my woodpile. (Everything is on the edge of a hill at my house.) I get behind the stack and put my shoulder to the bulge. It moves. I shove forward individual pieces that stick out too far. Some don't budge; others are less stubborn. Gradually, the whole stack reacquires a semblance of balance.

Theoretically, firewood is carefully laid, piece by piece, one on top of another. In actuality, when I can't reach the top of the back row over the top of the nearer row, I start throwing pieces of wood onto it. This is not really a good idea because theoretically the shock of the sudden weight could send the firewood tumbling downhill. In actuality, it sometimes does.

Theoretically, the whole structure gains in strength and solidity as it grows. In actuality, it's only when there are many, very heavy pieces of wood already in place that the wood goes crashing down the ends.

Theoretically, the whole pile could tumble, but in actuality, some new order asserts itself with some strange twist of a particular piece of wood, and the disintegration stops. The best thing to do is to laugh and build again. After all, it's only a real-life version of my father's Blockhead game, in which players take turns placing their blocks on the common structure until someone's block brings the whole thing crashing to the table. In the same spirit of games, taking pieces of firewood out of the pile thrown off the truck is like a game of pick-up sticks. In either Blockhead or pick-up sticks, whether on the living-room table or with firewood, it's best to laugh when a wrong move brings the structure tumbling. At least I've learned to never try to stop the roll of the disintegrating firewood stack with my foot. I politely wait till it's my turn.

When all the firewood is neatly, aesthetically stacked, I assess the difference between theory and practice. Theoretically, the black plastic is tightly enough battened down not to blow off in a rainstorm. Theoretically, there is enough firewood to last the winter because the rains have already made the road impassable. In actuality, I'm always afraid, every winter, that this will be the coldest winter in two hundred years and I will, as I've always feared, pick up the last piece of firewood from the woodpile in March and wonder how I will stay warm till spring. In actuality, it's always been just enough.

Goldberg Variations on the Mountain

One evening I went into Ashland for a concert of Angela Hewitt playing Bach's *Goldberg Variations*. To my delight, for the first half of the program Hewitt talked to the audience about the music; then she played the *Variations* without interruption, having particularly cautioned us not to cough in that crucial hanging moment at the very center of the piece. She sparkled in lavender lamé pants; her lecture sparkled with insight and joy, and the music - ah! bright wings! In this composition Bach painted all the variations of life - the lyric and the poetic; the sorrow and the pain (despair, Hewitt called it), the joy, the "romp on the green," the leaps and splashes, the solemn, the funny, the tender, and, at the end, the nostalgic remembering, like the beginning of *Fanny and Alexander* when, at the family Christmas gathering, the old matriarch and the old man walk arm in arm through the house, reminiscing, laughing now at the passions of their youth, when her husband had caught them in a lovers' embrace. It is a nostalgia that looks back on the joy and the pain, sighs with fullness, pats an arm, says, "It was, all told, such a beautiful thing, this life we've had."

Surrounding all the variations Bach built out of his opening theme was a sense of beauty, so much so, that, leaving the concert, I held within me for a fragile moment the entirety of life, all its depths, hollows, peaks, and fields connected by this sense of beauty. What I wasn't prepared for was life's ugliness. What an emotional shattering it was to stop at a gas station just outside Ashland. Lights blared and rock-and-roll crashed into my car window, flattening Bach with the punch of a thug. Neon yelled at me. "Bud Lite!" it shouted. "Open!" "Chicken and Jo-Jos!" "Beer & Wine!" "Hot Dogs!" "Marlborough! $1.95!" The open garage, as though under the harsh glare of operating room lights, exposed the grunge and hard metallic clutter of its insides. The odor of gas wavered in the air, drowning my nostrils.

At last, home on the mountain, I stepped out of the car and could hear again that Bachian beauty sing. The night was so dark the trees, black, gave light to the sky by contrast and there was no path up to my house, no earth, even, no separate trees, only a flat cardboard cut-out with jagged tree pinnacles and, beyond that, the universe in which stars danced white fire, giving distance. The frogs gave distance, too, with their far-and-near song, that splash of rivets as abruptly ended as one of Bach's variations. I walked up the hill and stood in the dark on the lawn, where the cardboard cut-out now shaped an outline of jagged trees, the line of a ridge, the upward curve of the mountains, over which stretched a black canvas of sky where God had flicked stars off the end of his brush like Jackson Pollock. The frogs were silent, but now the continuous rumble of the creek down the mountain gave distance - depth - to the sense of cardboard.

Bach must have understood stars and frogs, mountains, forests, and rivers. Without that understanding, the *Goldberg Variations* would be a cardboard cut-out. With it, Bach could give us all the emotions of a lifetime and then say, looking back (oh, that beautiful opening/closing theme!), "It was, after all, such a beautiful thing, this life."

For the Love of Light

For years during the long, dark evenings of winter, I have lit the main room of my house (the 10x12 "one-room cabin" with its stove, sink, writing desk, couch, and bedroom loft) with four kerosene lamps and one little glass oil lamp. I huddle under three lamps at my work space and keep another lit over the stove in the far corner of the room. I carry a lamp with me through the semi-dark when I move from one spot to another.

This year, fed up with the dimness of those winter evenings, I went on a binge. I wanted a glory of light in the house, light behind me as well as in front of me, light always there even when my back was turned. And so I lit candles, and lo! the light shone brightly. The entire room came alive with the bright breath of living fire. The soft, cheerful candlelight was as joyful as a baby's laugh; the gentle flames swayed slightly and flickered with life. The palpable essence of fire as light added a presence to the house that no electric light, bright and practical as it is, could ever produce. So I had light and its side-effect bonus of atmosphere, but it didn't last long. Candles are too expensive to use for light; such exorbitance is beyond my means.

Too enchanted with the effects, however, to give them up, I called on the tradition of Christmas giving and told all my family and friends that what I really wanted for Christmas this year was candles.

And so I got candles for Christmas, dozens of candles - thin tapers, little votives, tall fat candles and short fat ones, cheap candles and expensive, scented ones, hand-dipped candles of dense, subtle colors and commercial candles of almost transparent white, green, red, and blue. As I opened package after package of candles, exclaiming each time, "Yes! More candles!" people began to discreetly roll their eyes, suspicious of my enthusiasm. But they are those who live with electric lights. They don't understand.

The whole idea of using candles as a primary source of light was so odd people felt they had to warn me about it, too. My son told me he had read somewhere that candles were a major cause of air pollution in the house. I said that seemed of little concern in the face of kerosene lamps and wood-burning stoves. He agreed with that and added with a hastily suppressed smirk that my house was so drafty candle pollution wouldn't make any difference, anyway. Several people warned me of the danger of fire, which is true: cats' tails brushing across the open flames easily burn, filling my nose with the stench of singed hair.

I have used my Christmas candles exactly as I had planned – extravagantly, for brilliant, soft light. Every night I light the four kerosene lamps and the one oil lamp, then add to that, seven taper candles, two votives, and one long-burning fat candle. I go through more than a dozen candles a week. At this rate, I'll burn up all my Christmas candles long before winter is over. But it doesn't matter. I have loved the house in full candlelight, and when I return to the dim shadows of the four kerosene lamps and one oil lamp, my memory of the candlelit beauty of my house will keep a residual glow there for many months to come.

Rolling Blackout

I came home from town late one evening in March, threw a log on the fire, and turned on my propane stove to make dinner. Drat. I had run out of propane. I wasn't in any mood to walk back down the hill again in the dark and the cold just to change to a new five-gallon tank, so I decided to do it the next morning, when I had to walk down the hill anyway. I could cook on top of my wood stove tonight.

After eating dinner and doing chores, I took a book into my bedroom loft with me for some bedtime reading. I turned on my new bedside lamp, which runs on a powerful, rechargeable battery, and read for about five minutes when suddenly the light turned yellow. Drat. The battery was dead. I set my book aside and went to sleep.

Just before dawn (as I estimated), I woke up shivering slightly and thinking I would have slept better if I had thrown that extra blanket over me. The night had been colder than I had expected. But I figured I would have to get up soon, cold or no cold, and I turned on my flashlight to look at my clock. Two minutes past midnight!? Surely it was later than that! Was that really moonlight, not dawn's light, seeping through the skylights? I looked at the clock again. Its silent face stared unmoving back at me, and I realized that its battery had quit during the night.

I jumped out of bed, scampered down the ladder, and opened the stove door to restoke the fire on its bed of overnight coals – or so I would have done except this morning there were no coals. The fire had gone out. No wonder I had been so cold during the night.

As I went outside to chop some pitch wood to get a fire going quickly, I thought how lucky it was that the lamps still had kerosene; otherwise, I would have had no light at all in the house in addition to no heat and no propane for cooking and no bedside lamp.

When I threw the tarp aside to get to the firewood, I was surprised how frosty it was. The night had definitely been colder than I had anticipated.

Shivering as I came back inside, I quickly relaid the fire. As soon as it was flaming warmly, I stepped to the sink with my kettle, intending to heat up some water for tea. When I stuck the kettle under the spigot, though, and turned the faucet, no water spewed out. The pipes had frozen.

In the nineteenth century Oliver Wendell Holmes wrote a long poem about a shay built by a deacon who was determined that it would wear out before it ever broke down. Every piece – wheels, thills, floor, sills, panels, spring, axle, and hub - was as strong as every other piece. One day while the deacon was taking a drive, the whole thing fell apart. "You see, of course, if you're not a dunce," says the poet, "How it went to pieces all at once, -/All at once, and nothing first, -/Just as bubbles do when they burst."

That's just how I felt - as though my shay had fallen apart all at once - that day at my house when I had no gas, no light, no way to tell time, no fire, and no water – all at the same time. It was, as my sister said, a rolling blackout in a nonelectric house.

Rooted in Story

When I met the Irish storyteller Tom Foley, who lives in this area and travels around the country with his show, *A Celtic Christmas*, he asked me about the house I built. As the conversation progressed, however, I must have momentarily forgotten to whom I was talking, for at one point I said proudly that my living in this house for twenty-five years had given my son a valuable sense of place.

"I know what you mean," Tom answered evenly, without a trace of condescension; "I grew up in a house my family lived in for 200 years."

200 years!? With my quarter century of residency reduced to babyhood, all I could do was marvel, "Oh, what stories those walls could tell!" - which Tom thinks is true of more than walls. In western Ireland, he says, "not only did every person have a treasure-trove of stories; so did every place, every field, every scythe, every milk churn, every glen, meadow and bog." One of his purposes in bringing these stories to an audience is to reconnect us with an ancient sense of community. "Without stories there is no community," he says, and he believes that today we have lost both.

But most people today don't stay rooted long enough to build stories into the walls. If we have lost our sense of story, maybe it is because we have lost our sense of place. Without the place to anchor the stories - the same place, the same stories, generation after generation - the stories become unhinged; they float and vaporize like ghosts, or, like stuffed eagles in museums or rams' heads in hunters' homes, they reside in books and theatre performances. "It is the place that matters, the place at the heart of things," says Loren Eisley. "We cling to a time and a place because without them man is lost, not only man but life." Our attraction to Tom Foley's Irish stories is our yearning for that sense of place without which we are lost and which is at the heart of the stories which are at the heart of community.

But Tom's brother, at home in Teampal an Ghleantain, is building his family a new house. He doesn't want to live in the drafty old family house any more, and who can blame him? My parents still live in the house I grew up in, but when they die, I won't be moving back; my son lives in Seattle, and Tom is raising his son in Ashland, not in Teampal an Ghleantain. Our modern mobility doesn't give stories a chance to grow and then to settle onto every rock, every butter churn, every wall. Without a sense of place, stories can't grow; without stories we lose community. But we can't undo the ribbon of progressive history and keep people rooted generation after generation, as of old, so how can we keep our sense of story?

"We must begin where everything begins in human affairs," says writer and monk Thomas Berry, "with the basic story, our narrative of how things came to be, how they came to be as they are, and how the future can be given some satisfying direction. Interior articulation of its own reality is the immediate responsibility of every being."

For Berry the story of Earth and its emergence into the universe, the great scientific story of evolution, is the basic story that identifies us with place. As mobile populations make the Earth smaller, Earth itself is the place with which we must identify. Only when we understand the story of the Earth in terms of our own histories and place ourselves on that Earth with the connection of our own stories, can we find that interior articulation of our own reality that gives purpose and meaning to our lives.

Changing the Charm

Ever since I built my house many years ago on a very limited budget, the ceiling of the original house and the interior walls of the addition had remained as inadequately covered as they were when I ran out of money while I was building. Last summer I looked around me and said, "I want real walls." Friends warned me that putting new walls on my house might diminish its charm, but they weren't the ones living with a black plastic ceiling and walls of burlap or cardboard. I didn't think the charm came from burlap and cardboard, anyway. I was determined to put real walls on my house.

On my way to town to buy lumber, I stopped to see a friend and told him what I was doing. He said, "I have some old lumber behind the barn. If you like it, I'll give it to you." It was sixteen-foot 1x8's and 1x12's, milled off the land of mutual friends who had given the wood to Christopher when they moved from their land. It was old, pocked, red - perfect for my "mountain style" house.

That auspicious beginning was given the lie the next day when my friend Tom Cregan, bringing the wood from Christopher's place to mine, was backing his truck in a great roar up the hill to the bottom of the path to my house. Watching to avoid the ditch on his left, he rammed into the tree on his right. I was aghast; Tom smiled wryly at his crooked bumper; we unloaded the wood, Tom left, and my work began.

A sixteen-foot board is very, very long. Each one stuck out eight feet in front and eight feet behind as I carried it up the trail through the woods. "As stiff as a board" is not an inept metaphor; a board insinuates itself around nothing. The trail to my house, steeper with each consecutive trip, curves sinuously between trees. At one turn in the path, clearance is fifteen and seven-eighths inches. At another place the path squeezes board and carrier between two large firs. I could carry only one board at a time, now on the right shoulder, now on the left, now on the hip, and, at the end of the day, up the hill again and again with a board on the head.

The next day, while I was in town (and suffering a headache), my neighbor, Sylvia, carried more boards up the hill for me. What a blessing of a neighbor!

The fourth blessing in this project was my son, Ela, who, with a few precious days to spare, came from Seattle to help. First we tore the black plastic off the ceiling, which took a bit of nerve, as I expected scores of scorpions to fall out, but the only scorpion we saw that day fell out of my long-unused coffee pot when Ela opened it to make some coffee.

We tore down shelves, piled books and furniture in the middle of the room, ripped off burlap, screwed a ceiling into place, and started on the walls. But when I got up on the morning of the third day, my heart sank. The room with its two new walls of old wood looked very dark. It would be dismal in the winter, and I knew I would hate it. So I jumped in Ela's truck and went to town to buy new, light, tongue-in-groove cedar, which Ela and I put up as soon as I got back. The gods were smiling again. The room had brightened.

After four days Ela returned to Seattle, and I turned my attention to the upstairs walls. Oh, how I missed Ela's quick, careful work! And how I missed his electric tools, too. I was using a hammer where he had had a cordless drill, a carpenter's saw where he had had a Skil saw (that my little generator barely ran). But I can hammer, and I can saw, and the walls are going up, with the rest of the old wood and the new. Every night I crawl into bed exhausted, every muscle sore and my hammering arm aching. Every morning I get up and look at my walls, and I love the new look in my dear, old house, which has confirmed the auspicious beginning of this project many times over, for it has lost none of its charm.

A Flurry of Children at my House

When my friend Joan told me that her daughter was coming from Montana to visit with her six children and that her son and daughter-in-law, who live in Hawaii, would be here at the same time with their two children and then suggested that my house would make a good excursion for all the grandchildren, I took a deep breath. It's just a little house, but it was late summer and the weather was good, so the outdoors could handle the spill-over from the house, and I was flattered to think of my house as a tourist site for children, so I said sure, I would love to have them, and we decided they would come some time over the week-end.

Joan called at 1:00 on Saturday to say they would be over around 2:30. "I should make some cookies," I thought - or I could make a run over the mountain for rocks for garden terraces, as I had planned to do that day. My next thought was that all those children would make a mighty good work force for carrying rocks up the hill, so I hurriedly found a box of saltwater taffy in the pantry, dumped it into a candy dish, and jumped in the car to go after rocks.

An hour later I was carrying rocks from my car up the hill. I had just returned for a third load when my guests arrived, spilling out of their van like leaves in a whirlwind. One little girl handed me a basket of flowers and child-made cookies, then dashed after the other children, who had taken off up the hill. Joan was calling to them to stay on the path and watch out for poison oak, and I didn't have time to be subtle. "It would be great if you could carry rocks!" I yelled, and suddenly – leaves sucked in - kids were clustered around me at the back of my car, hustling for rocks. Even the littlest ones wanted rocks to carry, and the oldest girl and the ten-year-old twin boys had to prove their superiority by carrying two rocks each.

At the house, the children dropped their rocks and dispersed. Suddenly, the place was crawling with children, up and down ladders, in and out of doors (and windows), swarming around the house and yard like bees on a hive, picking things up, trying things out, asking questions. "What's this?" "How does it work?" "Can I pick an apple?"

"I have to go to the bathroom." "Can I mow the lawn?" "What's in this pouch?" "Can I carry more rocks up the hill?" (You bet!)

I took one boy into my bedroom loft to show him how the skylight opened. He immediately crawled through it to the edge of the roof and said hi to his mom on the lawn below. She was nonchalant, so I didn't worry, either, and drew in my head just in time to see a boy under the loft curiously turn over a glass-ball oil lamp ("What's this?"). I told him where to find soap to wash the oil off his hands as I climbed down the ladder to retrieve someone's half-eaten apple from the grapevine, passing, on the way, Joan's daughter nursing her baby in one room and her daughter-in-law running out the back door after her toddler. Joan was on the front deck, helping a boy find the juggler's beanbag he had accidentally thrown onto the roof of the bay.

Children wandered up, asked a question, walked off again. "Do you ever get lonely?" "What do you do during the day?" " Can I have a piece of this candy?" "How do you twirl a jack?" "Is this a flute?" "How do you play it?" "When I grow up, I want a house just like this one, except it'll have a big garden with sculptures and tumbling fountains."

Then, suddenly, they were gone.

In the ensuing quiet, I looked around and picked up two candy wrappers and one apple core; I wiped up the dab of oil that had spilled. The onslaught of children had left remarkably little impact – except for the large pile of rocks in my garden (every last one from my car) and the warm glow in my heart from an afternoon of delights with Haley, Levi, Macayen, Jasper, Justine, Braenn, Mahena, and Kainoa.

Stonehenge on My Mountain

In late December and early January the sun rises late in the day behind the firs and cedars in front of my house. It grovels low in the sky, in a narrow arc, and calls it a day long before I'm ready to do the same. On cloudless days, it shines brightly through my skylights and moves quickly across the floor of the house, spilling pools of bright warmth from west to east. On those days I want to cram everything I do into those short hours of sun – go for a walk, play my guitar with my chair in the sun, read on the couch where the sun pools, write at my desk with the sun at my back, wash the windows, turn the compost.

Slowly at first, and then more rapidly, the sun rises earlier and earlier, slipping farther and farther north each time, arcing higher and higher in the sky until on March 21 it peaks, as it happens, just over the top of Humpy Mountain directly to the east of my house. Humpy is my own Stonehenge. If I could, I would climb up there and erect two vast stones, taller than any tree on top of the mountain, through which the sun would pierce exactly at the equinox, spring and fall, into the window over my writing desk. Experts would marvel for centuries: Who put these stones here? How did they get them up the mountain? Why are they angled in just such a way? What is their purpose? But my house would have been long gone, and they would never think of a single woman sitting at her desk, watching the sun travel across the sky day after day, year after year, waiting for that moment of exultation twice a year when the sun struck through the stone into her house.

The gathering momentum of the sun's course in the autumn creates a sense of panic: Winter is coming! Am I ready? Is the firewood in; are the blankets on the beds? Are the windows tightly sealed and hats and mittens close to hand? After that day in September when the sun strikes its note through the stones on top of Humpy, the pace slows; the days darken as the sun sinks deeper in the sky, closer to the horizon in its narrowing arc that begins every day farther south. And the cold comes (and, if we're lucky, the rain and the snow); the animals hunker down, the plants tuck in, our panic abates, and the waiting sets in. On December 21 the sun hits the farthest point south it can travel; then it turns and heads again towards Humpy and the spring equinox.

The gathering momentum of the sun's course towards spring creates a sense of opening. We begin to stir into life slowly at first – those warm days of February, the pruning, the emerging of spring bulbs, but as the sun gallops towards the spring equinox, the senses awaken with increasing expansiveness. When the sun once again peaks climactically through the stones on the top of Humpy, we shout, "Hosanna; the world has risen again!" It is spring; it is Easter; it is equinox. Summer is just around the corner, and spring is i-cummen in: a host of golden daffodils, daisies pied and violets blue and lady-smocks all silver-white, thrushes' eggs that look little low heavens, and a sense, some days, that God's in his heaven; all's right with the world. From the equinox on, long days lie ahead, and we can pretend happiness is forever.

A Sense of Nature

26

A Synesthesia of Sorts

I've always been a little envious of synesthetics, people who see colors in the sound of musical notes or for whom each letter of the alphabet evokes a tone. What a voluptuously sensual world synesthetics inhabit when they can argue over the color of Monday or the taste of E-flat!

Most of us are not blessed – or, some may say, encumbered – with such a sensibility, but, maybe, if we want to, we can come close. I know that feet, for instance, can see, because of those nights when I have forgotten my flashlight and have had to walk up the hill to my house in the dark. When the night is moonless and a cloud cover has obscured even the dim aid of the stars, there is no demarcation in the forest between trees and air. The dark has a tactile density. Eye-blind, I turn seeing over to my feet. They look for the solid, hard earth of the trail, nosing against the uphill edge like a snuffling mouse. When they find the root step, I put out my left hand and touch the oak snag, smooth and barkless, where it ought to be; when they reach the first stone, I know to turn to my right and lean out of the way of overhanging oak limbs. My eyes are useless; my ears are of peripheral help. I am completely dependent on the touch of my feet, my eyes in the dark.

If I have experienced how feet can see, I have also experienced how eyes can hear. One morning I woke up before sunrise and lay in my bed in the loft looking through the open skylight into the forest below. As the sun came up, a beam slipped onto a long, thin strand of a spider's web that floated from the limb of a tall fir. The rise of warm air off the earth swayed the thread, and as it did, the sunbeam slid up and down the strand, playing it like a musical instrument. The rhythm was sweet and slow, bright with melody. I lay in my bed for a long time, listening to this spider web sonata.

The confluence of hearing and seeing occurred on another occasion when I saw a flock of eight geese flying over the fields of Provolt. Flying exactly, symmetrically in its place among the seven dark gray silhouettes, but barely distinguishable against the pearl-

27

grey clouds, was a white goose. Keeping up wingbeat for wingbeat in the rhythmic pulse of flight, it was like a negative of its neighbors, like a placeholder. It must be like that to have a beloved companion die: an emptiness in the shape of that person where that person had once always been. As the flock of geese flew silently southward, it was the accent of the white goose that gave the vee its patterns of music – the staccato within legato, the melody over the bass line, the high overtone whistle above the guttural chant of Tuva throat singers.

I called the rhythm of the sun on the spider web sweet, though I didn't taste anything, and I said I was listening to a sonata, though I didn't hear anything. I said my feet were my eyes on the path in the dark, though they couldn't really see, and when I picture again in my mind's eye the flock of geese with its albino accent, I think of music, though I didn't hear a sound. These experiences are too real to be metaphors. They must be a kind of synesthesia available to any of us who let our senses play freely with the objects they are given.

Hunting on the Edges of Things

The Applegate Core is a group of neighbors in the Applegate who meet once a month for a potluck dinner and, usually, some kind of interesting program. This particular evening, the gathering was at my friends', Joan and Christopher's, and the program sounded especially interesting: a bat expert from the Forest Service, David Clayton, would talk to us about bats and try to catch one for us at Joan and Christopher's pasture pond.

After dinner and just at dusk, David took us through the woods to the pond where he had spread a net for catching bats. He turned on his bat detector, a high-tech little machine that translates a bat's echo-locution blips into humanly audible sound, and talked about bats while we waited to snag one. A bat can catch up to 600 mosquitoes a night, he said. Many bat species are endangered because of loss of habitat (they nest in old snags) and a low reproduction rate (only one offspring a year, the lowest rate for any mammal that size). Bat wings, he told us enthusiastically, look like human hands, with fingers and a thumb. The bat detector chattered like static while bats swooped in the dark overhead, scooping up mosquitoes in those digited wings.

Occasionally David would shine the red beam of his flashlight into the net. Suddenly he sprang up and dashed to the water, saying he had one. Excitedly, he jumped into his round-tub rubber raft, which careened dangerously as he did, and paddled awkwardly but rapidly to the net. As he paddled I could see the kind of life he lived - at the dark edges of things: at the edge of silence, at the edge of the water, at the edge of the dark itself, waiting for bats to appear, eavesdropping on that sound no human should be hearing, and, hearing it, waiting on the edge of anxiety, like a native hunter. And when the bat was caught - the splash in the dark water, the excited dash to the trap, the red light on the prey, the gloved hands disentangling the bat, snarling mad, from the net.

My friend Apple is a bat scientist at the University of Washington. She studies the hearing systems of bats' brains, which, though unimaginably small, are much like the human brain and so could give insight into human otology. The complete bat skeleton Apple had laid onto a board like a relief sculpture looked so much like a human skeleton it was eerily recognizable. One night Apple took me to see her lab. Here, in a narrow room crowded with the high-tech tools of the modern scientist and lit with hospital-bright lights, she spends her days peering through a one-eyed tube of magnification at the crumb of a brain. Above us, on the roof, her bats cling to the wire walls of large, walk-in cages in their upside-down sleep under the stars. If one of Apple's bats is sick, she carries it in her armpit to keep it warm.

I didn't tell David about Apple, afraid he might think her work counter to his own, he building bat homes under bridges and earnestly trying to persuade the public that bats are good, she, with her microscope, looking and looking and looking at the brain of a bat. But they are both scientists, and they speak with similar tones of voice about bats. Apple would love to wait on the edges of things with David to catch a bat in a net. And I think David would like to visit Apple's laboratory. They would have much to talk about.

Being with Apple in her lab as she lovingly handled her bats and talked about her work, and then being with David at the pond as he spread the wing of the bat in his hand to show us its structure, I felt my understanding of the human relationship with animals shift into a certain perspective. One scientist catches bats to study ways to preserve the species, maybe for their own sake, but maybe because they are "good" creatures who catch "bad" bugs, and the other scientist studies the bat's hearing system to help people who suffer from hearing defects. I think, in the end, there isn't a whole lot of difference between these two people whose love for bats goes beyond their professions.

Nonetheless, were I to be a bat scientist, there is no question which kind I would be, I, who also like the edges of things.

Political Correctness in the Barnyard

In these days of political correctness, we must watch our tongues like hawks. It's an insult to call a woman a social butterfly or a man mousy. Why should the valiant mouse who, day after day, bravely faces cats, brooms, and worse, be the image of a weak, timid man? Why should the butterfly, working to keep life and limb together as she flits from flower to flower, be a metaphor for a silly socialite?

But the barnyard, I find, is badly abused in the vulgar tongue. We call a slyly spiteful person catty, and even though my cat Piñon does look sly when he tries to slink into my lap, he's not spiteful, and, anyway, my cat Rorschach isn't sly, so why should we perpetuate the stereotype with the adjective? On the other hand, skittish Rorschach exemplifies the origin of the derogatory term "scaredy-cat," but why malign Cat in such a way?

If you're a dog lover and think cats deserve this language, consider the dog metaphors. Something worn and shabby is dog-eared; one who is unreasonably obstinate is dogged; bad poetry is doggerel; someone who swims poorly is dog-paddling, and, worst of all, a world marked by ruthless self-interest is a dog-eat-dog world, and is that, I ask you, a fair image to originate from your favorite pet?

What is the lowest epithet we can spit on a man? "You worm!" we say with contempt and loathing. But what an outrage! Where would we be without the earthworm? This is not the Elizabethan Age, when all creation was placed on a hierarchical Great Chain of Being with angels at the top and worms at the bottom. This is the dawn of the twenty-first century, when we recognize the equal importance of all beings. How can we in the Ecological Age address a loathsome human as a worm?

The reputation of the whole bird kingdom is besmirched by one frequent adjective: birdbrain. What does this say about our national emblem, the eagle? Does the owl with his reputation of wisdom take lightly that epithet? Does the chickadee who has to find a safe place for her nest or the osprey who hunts the hidden fish or the grouse who must hide her eggs from the skunk? Could you with all your intelligence live as well? Should you, then, be called "bird-brain"?

We say that children who mindlessly repeat what has been taught them are parroting their teachers, but think how smart the parrot is to learn another tongue. How dare we malign this jungle beauty by treating him like a dodo, which is another slander. "You dodo!" one kid cries to another who is being stupid beyond comprehension, and yet, which of us was the dodo, since it was the human who drove the bird into extinction, surely an act stupid beyond comprehension?

It makes one wonder what, if they could talk with human language, the animals would use the human as metaphor for. When one creature was being unbearably, vulgarly, stupidly arrogant, would the animals shout, "You human!" Would Mother Fox complain, when Father Fox worked too hard, returning late to the lair night after night, that he was "driving himself human"? Would mother birds whose children soiled their nests chide, "You're humanizing your home"?

All this should make us feel a little sheepish, don't you think? What I'm suggesting, then, is a little political correctness in the barnyard so we can worm our way out of these metaphors .

The Deer Who Failed

In the bleak landscape of the eastern Oregon desert I came upon a shattered visage on the head of a deer carcass hanging over a barbed wire fence. This frown and wrinkled lip told a strange tale.

"Something spooked the herd," it said with its mournful silence. "We bounded away, leaping the fence. All but I. I misjudged my leap, and alas! you see me as I am now." The doe's inner groin had hit the fence, the sudden jolt throwing her haunches over the top wire and her body forward. The back feet, tucked for the leap, had caught in the second strand of wire, and there she hung, trapped, firmly woven into the fence. The pattern of her death was in the weaving.

She must have cried out in an uncharacteristic deer bleat of terror. Did her fawn trot anxiously around her, touching her nose, mutely begging her to release herself and leap away? Did the other does stand near for a while, their ears alert and twitching, then return to their browsing, moving unfettered across the desert and up the mountain? The fawn, in an agony of indecision between two safeties, would at the last moment have dashed after the herd.

Alone, the doe pushed frantically against her trap. The barbed wire clawed at her like talons. Again and again she pawed the substanceless air for purchase, then drooped over the fence, spent, frightened, disheartened, thirsty; then, desperate for release, she struggled again, but no matter how hard she kicked, she couldn't free herself from the death-weave of the fence. Little by little she lost strength. Despair replaced frenzy. Her head hung to the ground. The merciless desert sun hung in the sky until finally night brought a relief of coolness. The next day was slow to come, and by the time this second sun went down, was the doe so weary, despondent, and sunbeaten she had given up? Was she hanging there already like a corpse, draped over the fence, emaciation already staking its claim?

It was the thirst that killed her, dehydration, the great weapon of the desert, working its slow torture until at last, too weak to struggle, the doe let her weight sag from those gaunt haunches and passed into a semi-conscious, thirst-induced, hallucinatory state in

which she dreamed of putting her nose into a cool spring and drink-
ing her fill. Meanwhile, the hot, dry, relentless desert slowly sapped
her life, leaving her body, skin-taut and bone-lumpy, hanging on the
fence to dry until one day even that will be gone and only bones will
be left, woven in and out of the fence, to tell this tale, as though some
King of kings had said, "Look on my works, ye Mighty, and despair!"
And around the decay of that pitiful wreck "boundless and bare/The
lone and level sands stretch[ed] far away."

Wind on the Beaufort Scale

The Beaufort scale provides a measurement for the force of wind just as the Richter scale does for earthquakes. Beaufort begins his scale with zero for "calm" (air in which smoke rises vertically) and moves through "light air" (smoke drifts, but wind vanes don't move) to a "light breeze" at number 2, a wind that caresses the face and rustles leaves. Moving up in force, the scale goes from this "light breeze" to a "gentle breeze" ("gentle" in this case being stronger than "light"), which is defined not by leaves but by flags. A gentle breeze will "extend a small flag." Next comes a "moderate wind," which raises dust and loose paper and makes small branches move, and then a "fresh wind," which is not, as I had always used the term, one that smells good and relieves the stale monotony of a hot afternoon, but one in which small trees sway. If large branches are set in motion and the wind makes it difficult to use an umbrella, we have a "strong wind," number 6 on the scale and half way to off the chart.

A wind revved up beyond "strong" is a "near gale," in which whole trees are set in motion, a wind that's hard to walk against. A "near gale" turns into a "gale" when it starts breaking twigs off trees and impeding progress. A few weeks ago I woke up to find the back porch and all the woods strewn with green fir branches: a gale had hit the mountain.

One notch past "gale" is the "strong gale," a wind that "causes slight structural damage." Oh, so wait a minute. That same wind of several weeks ago tore the top off my stovepipe; were we experiencing not a gale but a strong gale, not an 8 but a 9 on the Beaufort scale, not 39-46 miles per hour but 47-54?

Be careful with your vocabulary; technically a "storm" is stronger than a "gale." In a storm, whole trees are uprooted. I saw two trees toppled last week, so our gale might really have been a storm, although loss of a stovepipe is hardly the "considerable damage" also used to identify a storm. At any rate, I know it wasn't a "violent storm," number 11 on the scale, because we didn't have "widespread damage," Beaufort's only identifying characteristic of this wind.

Finally, after the violent storm, is the "hurricane," measuring anywhere from 12 to 17 on the scale and blowing more than 72 miles an hour. Here Beaufort balks at description, leaving that space in the chart blank as though his imagination were not capable of defining the force of a hurricane. Nor is mine. Although I have experienced the Beaufort scale from calm to storm, I have never, here in my little house on the mountain, had cause to try to define a hurricane. And that's just fine. A "storm" is wind enough for me.

A Blue Moon by Any Other Name

Last Sunday night up over the mountain rose a blue moon. If it didn't look blue to you, you might not have known that it was the second full moon in January or that the second full moon of a month is called a blue moon.

Why, no one knows for sure, but maybe it has to do with a connection between the blues and the moon. When I was a teen-ager and in a mopey, listless mood, my mother would say, "Don't be so moony." When lovesick poets sigh and write poetry, they invoke the moon. So if the name of the second full moon were going to be a color, it would obviously be a blue moon, especially given the nudge of rhyme. A red moon or a yellow moon has no phonic pizzazz, and we couldn't call it a green moon because that should be the name of the moon's first showing after its dark phase. Mothers pointing to this barest glow of a sickle in the sky could say excitedly to their children, "Look! There's the green moon!" as they do now when the blue moon rises, and the children would squint at the green moon as they do at the blue moon, looking hard and puzzling harder but never admitting that, try as they might, the moon never looks blue or green.

But why should a month's second full moon be called by a color? We could call it "the moon of double vision" and recognize it as a time of great insight. Or we could call it "the latent twin," and when the second full moon of the month appeared, mothers could say to their children, "Look, dear, there's the latent twin," and children would grow up thinking latent meant "glowing" or "fully round" except for those with lexically astute parents, whose children would grow up understanding that one with latent talents would one day at last rise to fullness.

A blue moon is rare but not nearly so rare as a month with no full moon at all. This dolorous misfortune falls upon February this year, squeezed as it is between the blue moon on January 31 and the first full moon of March on March 2, for March, too, has a blue moon this year. Rarity has doubled: 1999 not only has two months with a blue moon but one month full-moon-less altogether. There will be no

February full-moon gatherings this year, no February full-moon cross-country ski trips, no women who run with the wolves howling their stories this February.

By analogy to "black hole" and by parallelism with "blue moon," we could call this phenomenon a "black moon." Once in a black moon would be even more rare than once in a blue moon, but there is no moon to call black, so we couldn't say to our children, "Look! There's the black moon," and the name fails to satisfy.

We can't look for a name from the languages of native peoples, who used lunar calendars, which by definition give every month a full moon, and English (or, as far as I know, any other language) never gave the phenomenon a name because, as one might say, it only happens once in a blue moon - once in two blue moons, actually - so the necessity for naming it never came up. So I get to play Adam. I could dub the phenomenon "the month between moons" or, maybe, "the month the wolves don't howl." Yes. We could tell our children that this February is "the month the wolves don't howl," and they'll know by that that February is a month without a full moon, and they'll be attune to the omission of the full round orb this month and its accompanying, eery, earthly music.

On Being Given a Hint
of the Mystique of the Land

*"If we have a wonderful sense of the divine, it is
because we live amid such awesome magnificence."*
Thomas Berry

A restlessness in the evening air drew me outdoors.
Crepuscular birds were singing stereophonically; trees lightly rustled
and tossed, and my long, thin skirt swirled like an eddy as I walked
through the woods. The sky was thickening with clouds petaled with
shifting greys - slate-grey over lavender-grey over charcoal in the
west and a purple-black orchid of a sky lowering behind Humpy
Mountain to the east. Thunder rumbled under its breath. A few drops
of rain hit my thighs through the thin cloth and left wet kisses on my
shoulders. I quickened my step; I would have to hurry to avoid a drenching.

A tip of a rainbow slid over the ridge next to Humpy. Surely
I could pause for a minute to behold a rainbow.

Like ribbons pulled out of a hat by a magician, the strips of
color lengthened into the sky over the ridge. With the mesmerizing
effect of a magic show, the strip of rainbow began mysteriously to
glow, at first gently, then more and more fiercely, sharpening in color
and splitting the sky between the heavy, dark grey on the outside of
the arch and an increasingly radiant embroilment of color on the
inside, where a bright lavender grey melted into a lighter, brighter
mauve that became at last a pulsating golden rose streaming to earth,
hiding the sky like a curtain, the rainbow like a wall separating the
black darkness from the unendurable brilliance; only the force of that
concave curve was keeping that blaze contained.

In the south, the rainbow's other end was being pulled
upward till the connecting center of the arch was traceable as a bent
strip of light across the sky, an achromatic moment before the arch
filled in with color and this luminous effulgence of the spectrum

divided the partial dome of visible sky into blackness above and light below, like an unknown hexagram of the I Ching: Li below, fire, brightness, beauty, but what was this above, this darkness in the sky as deep as night that was not night?

With a last wink of brilliance, the sun sank behind the mountain, and the hot fire inside the arch began to cool; the top of the arch faded and disappeared - the magician reversing his trick with exact symmetry. I gathered up my skirt and was turning to go home when a blazing, blue-white, horizontal, multi-forked streak of lightning cracked open the dark sky under the echo of the rainbow; it blazed like a mercury fuse, then closed down in darkness.

For us to enter the Ecological Age, says monk, scholar, writer, and wise man Thomas Berry, we must counter the industrial mystique with a mystique of the land, and he gives to the poets and natural history essayists the role of evoking that mystique. Let me say, then, that what I saw - the dark sky outside the rainbow against the glowing radiance on the inside, the brilliance of the colors, the constant continuum of their imperceptible changes, the beauty of the curve and its unmatched size, the peak of the mountain under that curve, the occasional thunder and spattering drops of rain, the dynamics of movement in the sky, the whip of lightning, a bird's song higher up the mountain like the wind against my skirt reminding me that it was here on earth, under what Berry calls the compassionate curve of space, that this was taking place - such glory suggests there is more beauty within that curve of space than we have yet seen and is a touch of something more than any god that puny we can postulate.

Cats and the Wild Things

Nepali Ama, an old woman who lives in the mountains of Nepal, once lamented about her life, "Hare Ram, I come home at night and who can I say I'm tired to; who can I say I'm hungry to?"

Except for my cats, I might be saying the same. Instead, I come home and say, "Cats? Hello, Cats!" They bound to me to be petted; they rub against my legs and welcome me home: Rorschach the disdainful, the scaredy-cat; Piñon the sweet one, the crippled one, with his funny sideways run.

I love my cats, but I don't think I have done a good thing by having them. They keep mice out of the house and moles out of the garden, but I think they are not good for the ecology of this tiny corner of the earth. There are two cats next door, there is a cat at the only other house on the road, and there is a wild cat, too, a great, ugly, mean, smelly cat who acts like my house is his whenever I'm not home. There are too many cats here, and cats are voracious predators.

Rorschach likes to sit on the roof of the pantry and snatch bats as they fly out of the eaves. She has to be quick to catch a bat, but she has been so successful that those bats she hasn't already caught have abandoned my house to seek their evening meals of mosquitoes somewhere else. It was not a good bargain. Rorschach does not also diminish the number of mosquitoes, nor can she imitate the swooping beauty of bats in flight.

Neither Rorschach nor Piñon catches many birds, thank goodness, but they do catch snakes, which makes me angry. When they drop the snakes on the ground, I chase the cats away and hurry the snakes into the woods or the grass. I cringe to think my cats might have depleted the snake population, too.

One night I found the long tail and fragile back leg of a baby coyote in my bed, which is Rorschach's favorite place to crunch her catch. Coyotes are literary heroes of many myths; they give us metaphors and symbols. They're not seen here often, and that my cat killed one saddens me unaccountably. I mourn the dead baby coyote and feel personally responsible.

And so I wonder if I should get rid of my cats and return the ecology to its proper balance, including, I suppose, mice in the house. I could come home at night and sit in silence in the doorjamb, waiting for the coyote to slink through the woods or the snakes to slither close enough for me to speak to. I speak to the birds and the chipmunks already; why is it I think I want cats to speak to, too?

Because, says the Fox to the Little Prince, "one only understands the things one tames. *Si tu veux un ami, apprivoise-moi.*" If you want a friend, tame me.

And so I have made my cats my friends. My friendship with the wild things, no less significant, is more distant. My social nature has needs that clash with my affinity with wildness, my enjoyment of my cats with my desire to live here on this mountain as part of its natural world. Domestic cats are exotics; they don't belong here, and there are too many of them. Like star thistles, they threaten to take over other wildlife.

And yet, when I come home at night, I can say to Rorschach I'm hungry, and she will run to her food dish in sympathy; I can say to Piñon I'm tired, and he'll curl into my lap and purr.

An Etymological Bouquet

I crave to know the names of things. Like Indians of old, like the wizards of Earthsea, I know the name opens the way to magic. So the flowers are my intimate friends because I know them by name. I call them by name. I say, "Good morning, you bright-eyed ox-eyed daisies."

It was the English who named the daisy - day's eye, the sun, as the daisy is with its yellow orb and white rays. As fresh as a daisy, one says, and I understand: they are so white, like my dress hanging over the deck railing to dry in the sun. It is appropriate that the daisy means purity in thought, innocence, loyal love. Loyal love? I pick a daisy to test that truth. Loves me, loves me not, loves me, loves me not...tension mounts. Loves me, loves me not, loves me! Hurrah! I pick daisy after daisy, deliriously desirous of hundreds of daisies so I can prove all day long he loves me.

I move from clump to clump until I am drawn beyond my usual path into sparse bushes - and new flowers. In England the hot rays of the sun are paled by London fog and English rain, so white daisies look like the day's eye. In Oregon the sunshine flower is all brilliant yellow, yellow orb and hot yellow rays, like southern Oregon's summer sun, and so the sunflower has been renamed Oregon sunshine. Gerard Manley Hopkins's thrush's eggs "look little low heavens"; my Oregon sunshine looks little low suns, galaxies of them. A dark purple cluster-lily growing between the Oregon sunshine strikes vividly against their gold. Unrelieved sunshine is too intense; the night must temper the day, the moon relieve the sun. Yellow suns and purple moons join my bouquet of day's eyes.

To the white, gold, and purple, I now add pink. Wild phlox splotches the woods with pale or vivid pinks. I am partial to the vivid varieties, and I leap from one group to another until I am deep under the firs that seldom see the sun, yet the colors of the phlox are still as bright and rich in one clump as they are pale and delicate in another. A bouquet of phlox is supposed to be a proposal of love, perhaps because phlox is the Greek word for flame. To whom shall I give my bouquet of phlox? To him who the daisy said loves me, of course.

I move out of the woods and into the warm morning sun, walking purposefully along the trail of the old mining ditch to a patch of Indian paintbrush. This is no ordinary paintbrush but a species endemic to the Applegate called Applegate Indian paintbrush. It took no great imagination to name these plants. Their flaming, scarlet-orange tips (bracts, not flowers), feathered into a fan, are botanical replicas of paintbrushes just dipped into a bucket of paint. I add some to my bouquet, where they contrast nicely with the delicate pinks of the phlox.

At the bottom of the path to my house, I stoop to observe the wild columbine, red and spiked, like a firecracker caught in explosion. Its spurs suggested dove heads in a circle to one imaginative pioneering botanist, so he named the flower "columbine" from the Latin word for dove. To someone else the same spurs suggested an eagle's claws, so the plant was also named "Aquilegia," from the Latin for eagle, so botany has given us, as it happens, the eagle and the dove in the same flower.

The columbine was once suggested as the national flower, the eagle being the national bird and the name columbine suggesting the District of Columbia, the nation's capital, but the matter was quietly dropped without much explanation. I have a suspicion it was because the columbine is a symbol of cuckoldry, a most unfortunate burden of symbology for a country to bear.

That's the same reason I hesitate to add the columbine to my bouquet. Would it belie the phlox and the daisy, cuckoldry giving the lie to love? The columbine is such a beautiful flower, though, that I am tempted to take it, anyway. Could that sense of temptation be the origin of the cuckoldry legend? Ignoring floral mythology, I add the bright red columbine to my pink phlox, white daisies, yellow Oregon sunshine, purple cluster lilies, and orange-red Indian paintbrush. Together, they make a bouquet of wild beauty and etymological stories that will brighten my writing desk for days.

Dear God the Weather

Applegate, OR
June 14, 2001

Dear God,

    With all due respect, sir, I would like to point out that there are better ways to conduct weather than those you have used in southern Oregon this spring. I know you have been doing this job for a long time, but could I suggest that you are very old and might suffer from an occasional "senior moment," such as forgetting how to put the seasons in their proper order? After a May that scorched us like July, you started June with weather like early April! It hardly seems necessary to tell you, of all people, Creator, that April comes before May and July comes after June and that the progression from winter to summer is one from cold to hot with a gradual and pleasant warming in between. That was the original, excellent design

    Maybe you thought you were making up for things by giving us the cool, almost-wet weather of the past few days, and, yes, it was a relief; don't think I'm not grateful, but, God, listen. Let me explain how it works. Winter storms put snow on the mountains, and during the warmer months the snow gradually melts and sinks down the mountain into the valleys, so everyone has water in the summer - the people, the animals, the plants. We need rain in the winter, when it's cold. June is too late! - too little rain - too warm for snow. Don't you know that? What do you think you're doing?! You're botching the whole job!!

    Please accept my apologies for losing my temper - and please don't get vindictive, though I know you can. We don't need any angry lightning bolts this summer - 'cause that's another thing. So little rain as you've given us this year has left us living in a tinder box. We're scared to death of fire this summer. Lightning causes fire, in case you've forgotten?

    I do sympathize with how busy you must be. Mistakes from the overworked should be forgiven, I guess. But, God, weather is an important job. If you can't handle it, or - if you don't mind my mentioning it again - if you feel you're too old for this job, why not turn it over to somebody else? I may not be the best person for the job, but even I would do it better than you. I'm sure you can find a good replacement if you're ready to retire. Wouldn't you like to vacation in the tropics without having to worry about the weather?

I would like to close with a word of thanks for all the beautiful things you have given us on this earth, especially here in southern Oregon. They are too numerous to mention. You have been generous and good to us in the past. Please don't stop now.

Give my love to all the angels. And in spite of my complaints above, much love to you, too.

Yours respectfully,
Diana Coogle

Nature's Little Gifts of Music

Nature is full of little gifts: the castanets of raindrops on leaves, the tympani-like THUMP-THUmp-thump-thump of grouse beating their wings against the ground, the warbling litany of raccoons, coo-oo-oo, through the woods at night, and the whistle of a wide-eyed doe, signaling caution, like a subdued train at a crossing, and giving the lie to the childhood myth that deer have no voice. Did those picture books with sheep that said, "Baaa" and dogs that went, "Bow-wow-wow" omit the voice of the deer because their authors hadn't heard it or because they couldn't find an orthography for this beautiful, whistled exclamation?

A walk through the woods releases music as though the foot, now here, now there, triggers a music box. Step, and set off a series of staccato chords: Chkt! Chkt! Chkt! - a frightened deer leaping away. Step again, and set off a sudden, loud, sustained crashing, like a Jimi Hendrix solo: a bear startled into escape. Step again, and by the subsequent suspirious sh-sh-sh-sh, like a thread pulled through the fingers, you'll know that a snake is slithering through salal. A lizard sounds like the same thing in rickrack.

Even a slight breeze will set the trees to talking, their creaks and squeals high along their trunks and branches, but windless days also have music. One hot, dry day last September, standing behind my house in a wood of mixed conifers and madrones, I heard a thin crackling, like a shower of very fine, broken, crystal glass or the bows of orchestra violinists bouncing pianissimo off the violin strings: ch-ch-ch-ch. If the crows' feet that break out around smiling eyes made sound, it would be like this.

Mysteriously, though, nothing was moving.

I broke through the veil of mystery when my eyes focused on the madrone trunks in front of me, with their layers of older, darker bark, shaggy and brittle, that had turned into curls as it dried and hardened, and the newly split, green bark beginning to curl back, exposing the beautiful red underneath. What I was hearing was the tinkling music of madrone bark splitting.

Would it be possible, then, to hear the whistle of a shooting star falling through the night? Could we, if the conditions were just right and we listened well enough, hear snowflakes as they fell or the song of the butterfly? How dare we say any creature is silent when the truth is only that we haven't heard it yet?

But it would be an unbearably noisy world if we could hear all the sounds nature makes, grass growing and flies mating and stars twinkling. Our selective hearing is necessary, but it leaves possibilities for more music than we usually hear. These sounds don't come to us as though we are in a concert hall, ready for the orchestra to begin. They come only when all the conditions are right and we are receptive and quiet enough to hear them. They are the little gifts of nature.

A Sense of Fun

Carberry Creek Dessert Bake-off

A few years ago when Ela was home from college for the summer, he told me one day that Linda, a neighbor of his dad's with a reputation as a good cook, had challenged me to a dessert bake-off. I'm not sure that that's exactly how it was; I suspect that Ela had simultaneously told Linda that I had challenged her. Surely the wily boy anticipated that both Linda and I, spurred by the spirit of competition, would spend the summer baking desserts and trying new recipes - and seeking the opinions of a willing taste tester.

We decided on five categories: a pecan pie, a fruit pie, a cake, a cheesecake, and an exotic. From then on, the summer was counted in dessert trials - banana sour-cream pie, Albanian walnut cake, Westhaven cake, leche flan, Moroccan date cake. "Your mocha pecan pie is awfully good," Ela would say, and then add slyly, "but boy, that pecan pie de luxe that Linda made yesterday!" and I was back into the cookbooks for new recipes: black bottom pie, cappuccino cheesecake, marshmallow-fudge brownies, peach tatin.

The morning of the bake-off I was in the kitchen by dawn. My sister Sharon, visiting from Georgia and enthusiastically participatory in the spirit of the competition, sat at my desk creating prizes and making calligraphy identification placards for each entry. Meanwhile, the kitchen was turning into a barely contained chaos of pecans shelled, limes squeezed, orange peels zested, ginger grated, and flour, sugar, eggs, milk, honey, and butter whipped, beaten, blended, folded, and stirred. The house contained culinary aromas like a gallery of tangible olfactory objects: long, wavering fingers of acidic orange peels; light prickles of spicy ginger; warm, homey clouds of baked pie crust .

Around midday Ela arrived from his dad's and jumped in to help. Linda's daughter, Sunshine, who had come with him, offered assistance, too, but Sharon said, "Don't trust her; she's a spy."

Just when the frenzy of cooking had eased enough for me to concentrate on the decorative details, I discovered disaster: the cat had helped herself to the topping of the sour cream orange cake cooling on the back porch. Sharon tried to reduce my panic by suggesting we could save it by shoving the topping around a bit, but nothing would satisfy me except that I make another whole cake, and so I did.

When we arrived with the five desserts at Ela's dad's and stepmother's home on Carberry Creek, where the contest was held, I, like Linda, was given a red hollyhock to wear. I pinned mine in the bosom of my white dress; Linda wore hers in her dark hair. Our desserts lay beautifully displayed on a white cloth down the center of the long table set up in the yard. Strawberries encircled the ginger brandy cheesecake; chocolate shavings dewed the whipped cream topping of the zuppa inglese, and a bright yellow marigold topped the new sour cream orange cake. Pinwheels of peach slices decorated Linda's peach pie; her eclairs swirled with chocolate, and purple pansies blossomed on the white icing of her lemon meringue cake. A large bouquet of flowers from Tracy's garden sat at one end of the table. Everything shone in late summer's evening light.

The guests - and judges - for the bake-off were the Carberry Creek neighbors, about twenty people, half of whom, my sister Sharon noted worriedly, were related to Linda. If family loyalty counted, I, flanked only by my son and my sister, was doomed.

Ela announced the rules. We would work by categories, he said. Linda and I would serve our respective desserts, which everyone was to rate on a one-to-ten scale. Ela would distribute and gather the ballots. Then we would move to the next category.

It all sounded organized and simple, but the ensuing hours were a kaleidoscopic chaos of milling people and beautiful desserts cut and served. The eager judges jostled in line, reached across the table, bunched up, held their plates out, said, "I only want a little piece," turned away, took a bite, said, "Mmm. That's delicious! Can I have another piece?" Linda and I cut desserts, filled plates, wiped spatulas and knives clean of whipped cream and sugar frosting, cut more pies, served more cakes. Ela's dad roamed through the crowd with his coffee pot.

"Coffee? Can I offer you some coffee?"

"Time for Category 3, exotics."

"Diana's zuppa inglese has got to be a ten."

"Have you tried Linda's eclairs yet?"

"Yeah. They're a ten, too."

"Do I have all the ballots in?"

"Coffee? Coffee?"

At first people felt a competition between friends was unfair and the judging part of the evening just an excuse for the event, but as the evening wore on, they got more serious. Linda was nervous. She served her desserts, then sat at a distance on the lawn, smoking a cigarette. I was a little nervous, too, but I was having too much fun to care a whole lot.

"Coffee?"

"Do you have any more of that peach pie?"

"I think the glazed plum cheesecake is one notch better than the ginger brandy cheesecake."

"Well, I think the lime tart wins for overall best dessert."

"Coffee?"

Finally, the table was a shambles. Empty plates smeared with whipped cream and littered with bits of chocolate and strawberry stems spilled over tables and benches; crumpled napkins lay in plates and on the ground; half-empty coffee cups grew cold in the evening air; forks cluttered the table like pick-up sticks, and the desserts were little more than crumbs and end pieces. The crowd was subdued, sprawled in lawn chairs and on benches and steps, replete, satiated, glutted, grinning.

Ela counted votes. In the end it looked like no one had paid much attention to who had made what, only to what they were eating, so the unfair advantage of Linda's large family was inconsequential. Hardly a point or two separated one dessert from another, and the winner kept looking like first one of us and then the other - until we got to the pecan pies. Linda's pecan pie scored higher than mine. Actually, it was better than mine. (So much for baking with honey to be more healthful, I thought ruefully.) And so the big bouquet of flowers and the first prize went to Linda.

I only cared for a brief minute. After all, a tie would have indicated that the judges weren't serious and would have rendered the prizes meaningless. We had all had an evening like none other, so Linda and I were both victors. And the following year, when Linda was dying of cancer, I was especially glad she had won the prestigious Carberry Creek Dessert Bake-off, which has gone down in the annals of history as one of the most magical evenings of all times.

Swimming in Crater Lake

It was already mid-afternoon when I arrived at Crater Lake National Park, and I was late. My friend who had invited – or challenged – me to meet him there for a swim was probably already at the lake, so I drove directly to the parking area above the boat dock and hiked down the trail in a quick fifteen minutes. I didn't see Alfredo anywhere. It was possible he wasn't coming, since he was on his way back to Portland, or maybe he had been there and left already. I went straight to the rest room to change into my bathing suit.

I had to wait in line for one of the bathrooms because there were a lot of tourists there who had hiked down the trail for the boat ride around the lake. The woman in front of me, friendly and chatty, as tourists often are, asked, "Have you taken the boat ride yet?"

"I came to swim," I said.

"Oh," she said, a little taken aback. Then, "They said at the lodge the water has 'warmed up' to fifty-five degrees." She gave a short laugh that was supposed to indicate irony.

I smiled.

I changed into my bathing suit, and then, since I still didn't see Alfredo, I climbed a good distance around the lake to get away from the people. I hopped, skittered, and climbed over and around rocks and boulders to a slight bulge in the roundness of Crater Lake, where a rock cliff fell straight into the water with dramatic beauty.

The tourist boats had shut down their motors. No one was in sight. I was alone on the deepest lake in North America, the seventh deepest lake in the world, sitting in a caldera filled with the best water God knows how to make. Above me on all sides rose the steep rocky peaks of the caved-in volcano. Snow was caught on the cliffs like snags of sheep's wool, and the cliffs themselves showed dark grey rock with soot-pink and lime-green streaks. Whitecaps broke the blue of the lake into polka dots.

I eased myself into the water.

After the first shock, it didn't seem very cold, and I swam towards the blue, away from the rock shallows towards the depths. The cold was not the problem. It was the wind-driven waves. What a battle it was to swim in those waves!

I came back to shore for my swimming mask and went out again. Oh, the exquisite underwater clarity - the reds, greens, and blacks of the rocks on the lake bottom and – just out there – the blue. That's where I wanted to swim – there, beyond where the bottom dropped out of Crater Lake, where there was nothing but blue, and so I pushed hard against the waves until suddenly the bottom was gone and I was swimming in the blue.

Visibility, surprisingly, was a problem because away from shore the water was suffused with millions of golden specks of pollen. I was entirely submerged, but the sun shining into the water reflected on each speck, and I was having a hard time concentrating on the blue. My vision was foreshortened, as though I were enclosed in the sky in a snowstorm. Suddenly, however, with a shift of perspective, I was looking *at* the dance of pollen instead of *through* it to the blue I wanted to be swimming in. Once having accepted the pollen as part of the experience instead of wishing it weren't there, I found myself enveloped by blue and gold. I lost direction. I had no sense of up or down, of surface or bottom as I turned and hovered and turned again. I was in space, floating, and all around me twinkled the galaxies of the Milky Way in a blue that was not earthly. If Crater Lake is not a crater but a caldera, for that moment, at least, it was also not a lake.

Backpacking in Mathematical Terms

Scott, a college mathematics professor and my good friend and annual hiking partner, pointed out to me as we hiked up to Horseshoe Lake in the Eagle Cap Wilderness Area of northeastern Oregon that decision-making can be cast in mathematical terms: 'A' equals the best choice if and only if the sum of regrets for not-A is less than the sum of regrets for not-B. Nonetheless, my decision to camp at Horseshoe Lake had been made by intuition, not mathematics.

The original plan was to camp at Six-mile Meadow and day-hike up the mountain to the lake, but when we got to the meadow, I suggested that we camp at Horseshoe Lake instead. Scott nodded, acknowledging the idea without commenting on its merit, but when pressed said if he were alone, he would certainly camp here, but this could be my decision. Remembering that snow had turned us back at that same altitude two days earlier (and influenced by Scott's desires), I agreed to camp in the meadow.

But my reluctance was evident, and when Scott said, "Now, you are the kind of person who would speak her mind, aren't you?" I said, "The problem is, I don't know my mind," and then, "Okay. I want to camp at the lake."

Immediately I knew I had made the right decision. Everything in my being was in harmony again as I hefted my pack and started up the trail, a sense of rightness so great it left no room for guilt for making Scott backpack when he had rather not.

Glistening at 7200 feet, Horseshoe Lake was large and dark blue, ringed by mountains and studded with rock shores. The lake sipped water from the snow at its banks; clumps of firs and cedars cooled the sun's heat in fits and starts. I had a superb swim. As we ate dinner in those magnificent surroundings, Scott said the only thing lacking was a tablecloth and champagne. We watched the sunset over the lake; we got up in the night to look at the stars swarming like bees; I had a long morning swim.

Thoroughly satisfied and happy, I said before we left, "This is, without question, my favorite environment." When Scott agreed that it was his as well, I looked at him curiously and asked why, if this were his favorite environment, hadn't he wanted to camp here?

Because, he said, backpacking was a way to get to places you can't otherwise get to. "Why carry a pack if you can get there without one?" he asked sensibly.

Because for me the purpose of backpacking is not just to see that place you can't get to otherwise, but to live in it, to know its shifts of moods according to time of day and kind of weather, to let the experience of the place sink into my subconscious. This, too, is mathematical. It is a function of time; its quality is multiplied by length of time, its intensity counted in exponential units of time. The sum of the difficulties of getting there - heavy packs, steep hills, sore feet, snowbound trails - is canceled if the arrival side of the equation is long enough and full enough. My decision to backpack to Horseshoe Lake wasn't based on thinking, "The regret of not swimming in the morning would be greater than the regret of arriving with a full pack to find camping inaccessible"; likewise when I was there, I didn't add up difficulties and pleasures to see which was greater. I only know that when I put my pack on my back and headed up the mountains, I knew I was doing right and that before I left the next day, I knew the equation was balanced.

Rumi in the Wilderness

In their book, *Backwoods Ethics*, Laura and Gay Waterman identify backpackers as more like Mae West or more like Twiggy: "Some people carry more than you can possibly use on a two-or-three-day weekend; others less than you need to enjoy a reasonably good time." Packing a backpack, I am of the Twiggy cast, and so, of course, books are left out. If I'm going to talk Twiggishly about cutting the handle off my toothbrush to save weight, logic dictates I won't be taking a book with me, even without its cover. Philosophy has dictated the same: In the wilderness, there should be no escape out of the present into the elsewhere of a book. Nature itself should be my entertainment and direct experience food for my senses.

The direct experience of my trip with Scott in the Mt. Jefferson Wilderness Area was to be tent-bound in a cold rain. Scott, who always backpacks with a book, was reading; I wrote a bit in my journal, then had no more to say. I fretted, bored. Scott read, happily. I cuddled next to him; he put his arm around me and kept reading. It hardly seemed fair to ask him to read to me; if I wanted to read, I should carry my own book. But where were my fine philosophies when I needed them? What happened to my reliance on the direct experiences of the senses? Drowned in the tiniest bit of rain that sent me scuttling to my tent! I am not, after all, merely a wild creature, who, content to be warm and dry, curls around herself and goes to sleep. A wolf's dreams are her bookish escape. But I am not a wolf, and I did not feel like sleeping.

"Scott," I pleaded, "read to me." Scott, always gracious, read to me. Together, now, we traipsed barefooted through the African village into school with Wole Soyinka, many miles and many decades from the cold rain and the tiny tent on the ridge-top in Oregon. I curled around myself, warm, dry, and happily elsewhere.

Since then I have always carried a book in my pack, if not the *Norton Anthology of English Literature*, then something small and light. On my trip into the Siskiyou Wilderness with my sister and her husband, it was a thin volume of Rumi's poems. As Billy fished, Sharon and I sat on the rocks at the edge of the Punchbowl, that 98-foot deep, 38-acre bowl of liquid sky, with our feet underwater, translated by transparency, the sun hot on our shoulders, the sky mirroring the lake, I reading Rumi to Sharon.

"I know I'm drunk when I start this ocean talk," Rumi says. "Would you like to see the moon split/in half with one throw?"

An osprey lazily circled the high blue ceiling over the lake.

"The sun hurries all night to be back for morning," Rumi says.

Clear water kissed our feet, turned turquoise just beyond our reach, built to a blue beyond the concept of blue deep in the lake where Rumi floated on the breath of my voice:

> I am sober now. Hand me my turban.
> Fill the skin jug, or give it back empty,
> Whichever.
>
> ...
>
> The moon and the evening star would dip down
>   like birds,
> If you threw the last of your wine into the air.

Change-of-Seasons Ritual

Sometimes, rare times, we can mark a change of seasons with such a perfect ritual that the satisfaction in the closure of one season is equal to the anticipation of the season to come. If we still want to be skiing when spring suns turn snow into heavy mush, we are out of sync; if we are lucky enough that the last ski is a perfect one, then it becomes a ritual of transition; we can put winter behind us with thorough satisfaction, pick up our shovels and hoes, and plant flowers.

Early in the day of a 15-mile ski trip at Crater Lake last week-end, someone skiing past me said, "It doesn't get any better than this." It was true. The weather was neither too bright nor too grey, neither too hot nor too cold, as striations of clouds jumbled over the sky like broken pottery. The snow was neither too icy nor too wet, hard enough underneath for easy gliding, soft enough on top for a good grip with the skis. The scenery was spectacular with ripples of dark mountains and snowy peaks to our left and the blue-gray lake nestled in a bowl of white cliffs to our right. Now we were skiing beside the lake, now past a cliff of horizontal slate suddenly cutting through the snow, each piece as carefully laid one on the other as though for a chimney; now we traversed a treacherous white slope with the steepness of an A-frame roof; now we crossed a windy field and stopped in a sheltered spot to sit on a cliff of snow and eat lunch under Crater Lake's tallest peak with its square lookout on top like a Shinto temple. Finally, cresting the pass, we beheld spread before us the round, white bulk of Mt. Bailey and, next to it, Mt. Thielsen exuberantly thrusting its King-Arthur's-sword top into the sky. The patch of Diamond Lake shone between the two mountains with just enough blue to distinguish it from the dense, white snow on the desert floor below us.

The climb to the top of the pass had been good - neither too strenuous for enjoyment nor so easy it was boring, but the long ski down the mountain was pure ecstasy, with expanses of uncluttered slopes steep enough for good speed but wide enough for good control, little rolls like sudden ocean swells for an occasional challenge, and clumps of trees here and there like obstacles in a board game. This effortless glide was like the making of a song on the mountain, an eagle-on-the-wing, dolphin-in-the-sea joy of unfettered motion in an abstraction of Nature's purity. It was like being alive in a photograph.

Driving home just before dark, up the familiar road through the familiar forest, I sensed something was different. Skiing at Crater Lake had served the purpose of a ritual, closing one season to make way for the next. The weather was fine here at home. My daffodils were yellow with full blooms, my plum tree raspberry-pink with buds; my pansies and anemones, newly planted in the flower boxes on my deck, flashed with color in the sun. Yesterday the first song of the returning birds accompanied dawn's opening. It is spring, and it doesn't get any better than this.

Thrills and Danger v. the Ancient Joy

Placid and quiet, the Sprague River flows through flat geo-
logical basins, old farms, summer homes, marshes, pastures, and
yarrow fields. Without a complaint or argument, with an easy current
that waves the water-maiden tresses of its underwater grasses, it
curves and meanders through the landscape like a twisted bonsai,
implacably itself. Gently, smoothly, continuously, for seven hours my
paddle stroked the water, sending the sleek red Au Sable canoe
streaming down the river.

Someone observed to me once the difference between hikers
as they come off the trail and rafters as they come off the river. The
former are individuals, each with an individual experience; the latter
has become a coherent group, people with a group identity. Canoeing,
I decided, creates a third sort, a grouping by twos, as canoeing partners, like rafting partners, forced to exist in a small space together and
to work together for a common good, become companions; yet still-
water canoeists, like hikers, have a more contemplative, quieter, more
internal experience than their white-water cousins.

White-water rafting, like downhill skiing, is thrills and
adventure. Still-water canoeing, like cross-country skiing, is peace
and contemplation, slow absorption of the countryside, a quiet in-
placeness with the river and its riparian community. In spite of being
a peace-loving, quiet-natured person, I like the thrills and dangers of
rafting. I don't always seek only quiet places; I need the excitement
of the rapids. Familiarity with danger, Melville said, makes a brave
man braver but less daring. It takes a sort of courage to face the big
rapids, and familiarity with danger, courage, I would develop.

But it takes a sort of understanding to canoe the gentle
rivers, a reaching towards depth, and understanding and depth I
would also develop. Sigurd Olson, observing the effect of the wilderness on the psyche, said, "As soon as we forget complexities and
problems, the ancient joy takes hold." It is that forgetting which is
important. Perhaps it is easier to forget those complexities and problems when faced with white-water thrills and the necessity of intense

awareness for survival, but I think that the ancient joy one discovers on the slower forgettings of still-water canoeing might be longer-lasting and make a deeper impression.

A Metaphysical Jaw-dropper

On one beautiful autumn day I hiked with a small group of Sierra Club leaders up the Metolius River in central Oregon. The banks lay almost level with the river, and though we weren't going appreciably uphill, the ice-clear water dashed past us as though pulled by galloping horses, running without sound over the hard-rock bottom like a stagecoach in a silent movie. The river wound through cedars and pines; in places flame-like viny maples pushed back the forest and hugged the river with a spectrum of yellow, orange, red, scarlet. Where the sun caught the leaves, they shone as though polished; where it caught the depth of the water, the river sparkled deep aqua.

We had been promised a hike that was a "jaw-dropper," but after about an hour of walking past such forests, such colors, such clarity of water, one of the hikers disagreed with the rating. "I've been to many places that are jaw-droppers," she said. "This isn't one."

I felt sorry for the proud Oregonian who had brought us on this hike, who had spoken eloquently and enthusiastically about this, one of his favorite places on earth. He admitted now that maybe he had overrated this place and exaggerated the beauty of this hike. But I didn't think so. Jaw-dropping scenery can be such that your jaw does actually drop open with amazement. You are astounded to see anything so beautiful, so grand, so sublime – Niagara Falls, Yosemite Valley, Crater Lake, the Alps at sunset. Other places have a quieter beauty that drops open your metaphysical if not your physical jaw. My metaphysical jaw was agape. The brilliant display of reds and oranges next to the deep aqua of the river was enough by itself, but if you add the dark cedar forest and the perfectly, unimaginably clear water racing with inexplicable speed over a nakedly visible bottom, if you add the blue sky and the open places in the forest where the land made a rounded elbow for the river to flow around, if you add the logs floating next to shore with faded flowers and laced with greenery, floating gardens that are one of the specialties of the Metolius - surely it is enough to drop the jaw of one who sees.

At the end of the hike I was reluctant to exchange earth, water, and feet for the metal and speed of my car, so it is no wonder that when I saw a sign to a turn-off for the Metolius Spring not too far up the road, I instantly gave up metal and speed to park and walk a long asphalt strip through gently hilled, wooded country to behold the source of the Metolius River.

The Metolius is not born small; cast immediately into adulthood, it begins like the river it becomes: big and fast. It tumbles out of the dirt at the bottom of the hill as transparent as a new dawn and galloping fast, as though Peneus, the god of rivers, were underground snapping a whip over its back, driving it from behind: "Go! Go! Go!" The river streams forward like progress itself. Leaping out of the shade of the forested hill into an open meadow, it sparkles in the sun, then dashes around a bend and disappears into distant hills. The long neon-green grasses in the river's first burst of life strain with the current like dogs on leashes. A full-sized fish flashes darkly among them; a pair of ducks bob their heads in and out of the water at the edge of the meadow. An eagle soars over the river, turns, and rises higher. At the end of the scene Mt. Jefferson shines white with new snow against the blue sky .

But being there at the source of the Metolius River, I was also not there. By the bizarre concept of ownership of land, a pretty wooden fence kept me absent. In case I might be tempted to duck through the fence and step into the river at its moment of birth, a second, barbed-wire fence behind the wooden one made a stronger statement. Here, it told me, I could look all I wanted, but if the grasses were whispering in the current, I could not hear them, and if the ooze at the edge of the source was emitting a gentle odor, I could not smell it, and how cold the water was my skin would never know.

And so I looked and looked, absorbing the scene by sight, imprinting it the best way I could, as a living, eco-sensitive painting framed by fences.

Nothing To Do But...

Here at the coast, I thought as I walked along Port Orford's ocean palisades, one has nothing to do but watch the tides go in and out or catch the sun sinking day after day into the vasty deep or listen to the surf whisper and cough and breathe its watery suspirations.

One has nothing to do but sit on a bench on a cliff above the ocean and ponder the mysteries of the universe in that wilderness of green with its occasional jeweled tips of white. One has nothing to do but smell the yellow, lavender, and purple lupines on the sandy hills or the ferns of the steep cliffs dropping to the ocean or the salt air and fish-and-seaweed aroma of the seaside walk.

One has nothing to do but walk endlessly along the beach, where no bend entices the curious walker to go just around the next one before turning back, where no trees and bushes obscure the view and hide the creatures, where one can walk and walk and walk, endlessly on sand, looking at the textures, colors, and patterns of pebbles and shells; picking up an unusually twisted piece of driftwood for the garden; watching the changing and eternally unchanging waves that carry an occasional bright green sea lettuce, dark brown kelp, or long hollow tube of bull kelp.

One has nothing to do but explore the tide pools in the crevices of rocks, where sea creatures wait - the bright orange, burnt sienna, umber, lavender, dark purple starfish stuck on the rocks like playdough figures in kindergarten, lumped together or spread out in their star patterns; the green anemones with their gaping mouths and beautiful flowing petals; the mussels packed onto the rocks like petrified wisteria blossoms. The water washes onto the rocks and over all the creatures and flows into the hollows and washes out again, back into the ocean, and the seaweeds wave and the anemones' tentacles flutter, and the starfish and the mussels never move. Another wave comes and goes and then another and another. Later, the whole rock will be under water, and, later still, it will be left even drier than it is now, and which of these creatures will wash back to the sea and which will stay on this rock?

In the fog of an early morning, two white plastic buckets stand out starkly on the beach. Beyond them, at the ocean's edge, barely discernible in the misty grays of water and fog, two fishermen in gray slickers cast their lines into the surf.

Priestess of Massage

Like a virgin of an ancient tribe preparing for some rite of passage, I stepped out of my sandals, took off my clothes, and slipped the earrings out of my ears. Naked, I lay face down on the table, one soft cotton cloth under me, one over me. Waiting, I let the sweet individual notes of a guitar settle around me, felt cello drift in, then leave, then come again. I breathed. I waited.

A presence glided in, the female shaman, the priestess of rites. Hands hovered over my back, then passed swiftly over my body. Lifting the cloth off my shoulders, they came down with gentle strength through bare skin into muscle, then deeper, even deeper, until they reached the psyche itself.

With those hands the priestess rubbed and stroked. She pinched and pulled and twisted. She bore down and let up and bore down again. She fingered and thumbed and palmed. She cupped her hands to fill pockets of warmth, comfort, ease. She heeled her hand to push into muscle. She kneaded muscle and plaited muscle and beseeched and pleaded with muscle. She palpated for soreness; she begged for release. With strong fingertips she struck points of tension that wouldn't let go. Pushing in, she pulled out, as on vibrating wires, that tension. Tiredness, strain, anxiety wavered up the wire from deep in the body and in those pinpoints of touch escaped. She squeezed and teased muscles into compliance till those hard stones turned to malleable clay with which she sculpted a riverbed through which began to flow the notes of the guitar and the voice of the cello.

She was the body's master; if it were to move, she would move it. She picked up limbs and shook them like dusting blankets; like blankets they lay where she left them. She rolled a shoulder - ah! massage that joint! She played the scale in slow notes of release up the arm, and the arm sang out with relief. She touched the face, and face said, "Yes! I am face!" Nose, eyes, cheeks, ears tingled with the anticipation of recognition. She touched neck, and neck lifted up, reaching for touch, pushing into it as a cat pushes against the hand that pets it. She touched toes, and toes twinkled. "Yes! Toes here!" She touched calves, and calves sighed, "Ahhh!"

The hands hovered again, then ceased. Well massaged into the body's version of nirvana, the body lay quiet. The priestess glided out. The rites were over.

The Evolution of Prepositions into Verbs

"Off," "on," "over," "out" - such words are not verbs, but in modern English usage they have subtly shifted from their old task as prepositions to a new one as verbs. It's a sort of evolutionary shift, like a species finding a niche and developing into a new species.

For instance, "to take a jacket" is not the same as "to take off a jacket," in which phrase "off" is not a preposition, as it is in "take the bug off the jacket," and if you "take off with the jacket," we're back to the original phrase with some important linguistic evolutionary changes.

In their new niche, these words act like verbs and are often intuitively understood that way. Misspellers, I've noticed, even if they aren't versed in verb-hunting, often hyphenate the verb and its adjunct, incorrectly, since hyphenation creates nouns and adjectives, not verbs. You can hold up a bank unhyphenated, in which case you have committed a hold-up hyphenated, but if that check you lost finally turns up, it isn't called a turn-up. You can have a check-up, a build-up, a touch-up, or some make-up, but there's no such thing as a turn-up - unless you slur your vowels and eat it.

These distinctions are tricky. How do you explain to students the philological difference in "up" between "He marched up the stairs" and "He looked up the answer"?

Some verbs are meaningless without their adjuncts. When you say you'll "carry out the plan," you don't mean you'll "carry the plan," lugging it in your arms up hill and down. You would say, "My hair is curled by the humidity," but "My cat is curled up on the sofa," by which you would not mean curled up as opposed to curled down. In fact, in visual-linguistic contradiction, curling up is like closing down, whereas uncurling is like opening up.

It makes sense to catch up but to fall down, to turn up a clue but to turn down an opportunity, to cheer up but to melt down. Although "up" and "down" are clearly opposites, "wash up" (hands, face) is hardly the opposite of "wash down" (walls, counters), and "to slow down" means the same thing as "to slow up." If I throw the ball up in the air, that's not the same thing as throwing my arms up in dismay, which is not the same thing as throwing up my dinner in disgust.

In other cases the addition of an adjunct radically changes the meaning of the verb itself. If a man says, "I knocked her down," that's far different from saying, "I knocked her up." A man might tell his children, "It's time to settle down now," but he would tell his poker partner, "It's time to settle up." As for boy friends, I might very well turn down the bad penny that turned up because we had a showdown after he showed up. Or I might hope we would break up before I have a break-down.

Some verbs change meanings according to their adjuncts. I might play up to a buyer while playing down my involvement in the scandal that implied I was playing around - at least until that story is played out. Or, look at this scenario: The boss tells you a woman is interested in signing the deal and asks you to feel her out. You would be in bad trouble if you changed the verb and started instead to feel her up. On the other hand, a change of position of the verb's tag-along may earn you a raise if the boss tells you about a certain deal and suggests you "see it through," but you, in your perspicacity, do not close on the deal because you can "see through it."

I could run on about this new linguistic niche, but maybe I should just run you off a copy. Okay, okay, okay. If you feel like you want to tell me to shut up, just hold off because I'm shutting down with my usual sign-off: This is Diana Coogle, from the mountains above the Applegate River of southern Oregon.

Language Parrot

Before I left for a visit to Sweden, I was delighted to find at a used book store a cassette tape for learning Swedish phrases. Although Maren and Lasse, the friends I would be staying with, were thoroughly conversant in English, I thought it would be polite to learn a bit of their language.

But parroting Listen-and-Learn Swedish is a strange way to learn the language. "Where is the cigar store?" is surely a useless phrase for me. "I want some fried chicken [pork, duck, goose, liver]" is just as useless, since I'm vegetarian, and, given the state of my pocketbook, I shouldn't have bothered to learn "I would like to go to the opera [concert, theater]," either. I couldn't imagine wanting to buy gold cufflinks, a silver compact, or a silk umbrella, but I would be able to if I needed to. I thought "What's the matter?" might come in handy, so I memorized it well, but then I wondered when I might use it: a small girl crying in the park, for instance. I give her a hug and say in my best Listen-and-Learn Swedish, "*Vad är det fragon om?*" and she, having understood me very well, pours forth a volley of tearful Swedish describing the meanness of her friend who left her to play with somebody else or telling me she is lost or that her mother has abandoned her, and I am as helpless as if I had never learned my Swedish phrases.

I learned the general useful phrases ("Where is the train station?" "Thank you very much," "How much does it cost?"), but Listen-and-Learn Swedish also provided me with phrases I hoped never to have to use - "I do not want the tooth extracted," for instance, or "The weather is bad." I thought I should learn "What are you doing tonight?" and "May I have this dance?" so I could understand them just in case they were said to me.

Was it fair to extrapolate something about the country from the phrases I was learning? These sentences from the "restaurant" section gave me pause: "I did not order this," "This is too cold," "This is not clean," and "There is a mistake in the bill." To be fair, Listen-and-Learn Swedish also provided "The food and the service were excellent!" but, still, it did make me think twice when I considered going out to eat in Sweden.

On the other hand, the tape didn't have any phrases like "The mosquitoes are terrible!" "There is a strike on," or "The toilet is stopped up." That was comforting.

The principle of substitution made me able to say many things other than the phrases I robotically repeated. After all, I am a thinking human being, not an automaton, so I could, if I wanted, say in Swedish, "I want to buy a train" or "Does this subway stop near a gas station?" If the "automobile travel" section gave me "Please lubricate the car," I could also say, "Please lubricate the railroad station." If I learned "My eyeglasses are broken" from the "difficulties and repairs" section and "Bring me a clean plate" from the "restaurant" section, then I could also say, by substitution, "Bring me a clean eyeglasses" and "My plate are broken."

Probably, though, when I get to Sweden, I'll be tongue-tied by shyness and won't say a word. I'm sure the Swedes would rather speak their good English than hear me stumble through sentences of bad Swedish, anyway. But I like entering a new language. It's part of the adventure of traveling. So, it's *Lycklig resa* – bon voyage – for me, whether anyone can understand me or not.

Coogleisms

Do I appreciate the country-born linguistic inheritance my parents have given me? Well, I should hope to kiss a pig! Nothing suits like the right phrase at the right moment, and my parents have provided me with a goodly share. When Ela used to say, "Mother, I need two dollars for...," I knew just what to say: "Ye gods and little catfishes. Another two dollars!" When I saw the living room floor spread end to end with bicycle parts, my mother's words would rise to my lips like apples in a water barrel: "This room looks like it's fixin' to walk away!"

On those rare occasions when I made Mexican pepper casserole for dinner, which I liked but Ela didn't, if he complained, I said, merely, as my mother before me, "You're not the only fish in the sea." That settled that as well in this generation as it did in the last. When I was a child and sick, my tightly-locked lips, confronted with a detested spoonful of milk of magnesia, would be gently prodded open by my mother's insistent phrase - her mother's before hers - "It's not what you want that does you good but what you get." The same phrase did occasionally prompt Ela to drink his healing teas, though it worked better before he entered the stage of not caring, if he didn't want a thing, whether it did him good or not.

Thanks to my father, I know the right phrase for ending a good meal: "I feel a lot more like I do now than I did when I sat down." As a child I thought I also knew the right phrase for asking for a second piece of dessert. "Just a little piece, like the first one," my father would say, and everyone would laugh, but when at a neighbor's dinner, I held out my plate for more pie and repeated his phrase, my parents were illogically more embarrassed than amused.

"Up and at 'em, Eve" roused sleepy daughters out of sleeping bags on camping trips, and "Hold the phone!" stopped cold any motion in the process of becoming. "Who do you think you are, the Queen of Sheba?" would make a haughty daughter slink away in shame, and what was a daughter to think when her father said to her, "You're my p-i-double-l pal"?

74

Dangling a child upside-down by the ankles, my father would threaten severe punishment. "I'll frap a hurtin' on your punkin head," he'd growl, and shrieks of terrified laughter would rise up from the floor. Years later when I managed to untangle the words from the sounds, I finally heard what it was my father was threatening - to make my pumpkin head hurt by what? By frapping it?! Blow me down, if I didn't actually find "frap" in the dictionary, meaning, in British dialect, to strike or beat. Now, where did my father acquire a phrase like that? From some distant Scots-Irish ancestor maybe? And it has come down how many Coogle generations?

These are Coogleisms, whether they came from my country heritage or out of my folks' own linguistic imaginations. Some of them I know came three generations to me, maybe more, maybe all of them, and four generations or more later they enliven Ela's language as well. "Good night, Ela," I would say as I put him to bed; "I'll see you in the morning," to which he would reply, as I to my mother, "Not if I see you first." Impertinent child. I should have threatened to frap a hurtin' on his punkin head and then told him, "That's for nothing. Now watch your step." That's the way to knock good country sense into a child.

Discovering the Difference between Labor and Leisure

I woke up and stretched lazily. It was Labor Day, a holiday, a day in which to do as I liked. What would I do? Soon I found myself facing my computer, fingers busy, as I contemplated Labor Day.

According to my dictionary, labor is a sort of work that usually connotes a fatiguing or onerous nature. It is less fatiguing or prolonged than toil, however, less painful or difficult than travail, and less demeaning than drudgery.

"Labor is doing what one must; leisure is doing what we like," said G. B. Shaw by dictionary example of definition. Was this writing of mine labor or leisure? If I chose on my holiday to write, then I must be doing what I liked, so writing must be leisure. My dictionary says leisure is "freedom or spare time provided by...a temporary exemption from work or duties," in other words, a holiday. Working backwards, I had discovered that holiday equals leisure equals writing.

But there was something about writing that seemed antithetical to leisure. This was hard work. This sweat of the brow couldn't possibly be leisure. I was back to labor. There I found colleagues illustrating by example the meaning of labor. Victor Hugo: "Thought is the labor of the intellect, reverie its pleasure." Carl van Doren: "With enormous labors he made himself into a popular writer." (I believe that.) T. S. Eliot: "The labor of sifting, combining, constructing, expunging, correcting, testing: this frightful toil is as much critical as creative." I toasted with Hugo, van Doren, and Eliot the labor, even the toil, of writing. Then I found Ellen Glasgow: "The act of scrupulous revision (endless pruning or trimming for the sake of a sound and flexible prose style)...provides a writer's best solace even while it makes drudgery."

Drudgery?! Was I going to spend my holiday in drudgery? What happened to leisure, reverie, pleasure? This was a holiday. Could I write today or not? I asked the dictionary, and it said, "A holiday is a day marked by a general cessation from work as an act of public commemoration of some event," in this case the commemoration of labor. So, in order to commemorate labor we are supposed to cease labor. The dictionary, with its helpful examples of words in use, has made it clear that writing is labor, can even be toil and drudgery. This labor was no holiday. Abruptly I left the computer.

Now I had lots of free time and was full of energy. I had been too busy lately to haul and stack firewood; now would be a good time to do that. It felt good to be working on my firewood early in September, building that mound of compressed, chemically different fire into usable form at my back door. Wait a minute. Working? On a holiday? There was the old contradiction again: work (etymologically kin to wrought and overwrought) and labor (with its fearful offspring, collapse) as opposed to leisure (etymologically connected with pleasure). I felt stuck in those antonyms. Who would help? Robert Frost, maybe, with his sense of uniting avocation and vocation, considering his work a sort of play for mortal stakes, but Yeats even more:

> Labor is blossoming or dancing where
> The body is not bruised to pleasure soul,
> Nor beauty born out of its own despair,
> Nor blear-eyed wisdom out of midnight oil.

It is not that I need oppose thought to reverie or labor to leisure, I realized, but that I find a life in which labor blossoms and dances.

Jungian Shadow Holiday

The origins of Halloween customs are mysteriously misted, appropriately enough, but we know the name comes from All Hallows' Eve, a date on the calendar of the Medieval Church. Some customs, however, come from ancient Celtic rites connected with Samhain, Celtic lord of death; some come from Medieval beliefs in ghosts and witches and some from Roman festivals associated with the harvest. Thus, we bob for apples, roast chestnuts, and eat pumpkin pies. As for the candlestick in the pumpkin shell, Irish legend tells about Jack, an old miser who died and, barred from heaven and barred from hell, was condemned to wander the earth with a lantern till Judgement Day: Jack of the lantern, Jack o' lantern.

But, above all, Halloween means costumes. Obscurity clouds the origins of this custom, too, but it could derive from a Medieval custom of the populace parading around the church on All Hallows' Eve costumed as saints, angels, or even (for the daring) devils. And because spirits are abroad on Halloween night, witches stalk and ghosts wander, but in this day and age, any kind of character could lurk in the shadows. For instance:

Once I went to a party costumed as my own grandmother, a disguise so effective a neighboring child asked me politely the next day if my grandmother were still visiting. Another time I was Lady MacBeth, wandering from guest to guest at the party, wringing my hands and crying, "Out, damned spot! Out, I say!" Once I was the pope, casting blessings in Latin, and once a witch, mumbling curses over potions. Another year I went to two different Halloween parties, first as a straight-laced prude and next as a harlot. My conclusion was that harlots have more fun than prudes.

The next year I became a fully bearded, royally robed king. That was during my years of poverty, when I had to rely on food stamps to make ends meet, and I had worn my costume to a function in town, forgetting that I would also have to collect my food stamps that day. Ignoring the amused glances from clients and social workers, I stepped up to the window to sign in for my appointment.

Shaking my head in dismay, I said to the woman behind the desk, "It's a sad state of affairs when the king has to apply for food stamps."

One year, when my son was very young, he and I went to a party as parrot and pirate, he, parrot, perched on my, pirate's, shoulder, at least for the costume parade. I once taught a class on Halloween day posing as a bear, a costume I hope none of my students thought appropriate to my mood. Last year, dressed in rags and carrying a basket of matches, I wandered up and down Ashland's main street as the Little Match Girl, crying out in the cold and asking strangers to take pity on a poor little match girl and buy some matches. I received my comeuppance, though, and learned not to approach people who might not be familiar with Anderson's tale, for one young man, a foreigner, mistook my disguise for the real thing and dropped a quarter in my basket. When I laughingly tried to return it, he realized his mistake and walked away angry while I turned more soberly in the other direction with a clearer insight into societies where the cold and hunger of beggars are much more real than a Halloween trick.

There is a refreshing cleansing of the soul in this expression of one's Jungian shadow on Halloween. Without Jack o' lanterns and costumed ghouls our unconscious selves would rise to power and devour us, or, just as bad, we would become robots of the middle road, unable to feel, to express. I tend to metamorphose on Halloween, and you'd never recognize me, but look closely this Halloween. You might see me on my broomstick, wrapped in shrouds and swigging the brew of hobgoblins.

Meeting Scrooge

I was walking through town the other day, humming the tune from "The Little Drummer Boy" and ticking off in my head the exciting list of things I had to do for Christmas, when I ran into Tom, Dick, and Mary Scrooge. "This could be a sour note in a merry day," I thought but stopped to say hello anyway because, after all, it is Christmas.

"Hello," I said, "and merry Christmas!"

"I hate Christmas," they said, as I had expected.

"That's too bad," I sympathized. "I love Christmas."

"It's so over-commercialized," they whined in their tiresome clichés. "I hate muzak Christmas carols and all the Santa Clauses that say, 'Buy, buy, buy,' - plastic wreaths, plastic bells, plastic toys. I hate the mall - it's so crowded. It's so expensive. I hate the obligation of buying presents," and they told me a story they had read in the paper about a Japanese visitor to America who said he couldn't understand this month-long celebration of shopping.

I didn't really want to argue (let them be happy in their misery, I thought), but it is the season of helping those less fortunate than ourselves, which they are, so I tried.

"Stay out of the malls," I suggested. "Don't buy gifts."

"Yeah?" they sneered. "And not give anybody anything? We can't do that."

"You could if you wanted to," I pointed out, "but what I meant was to make gifts instead of buying them."

Tom, Dick, and Mary Scrooge gave a short merriless laugh. "We don't know how to make anything," they objected, and I said, "I didn't know how to make hammocks until I decided to give them for Christmas one year. Or try reading books on tapes. That's hardly a crafts skill. All you need is a little imagination. It's ever so much more fun to make gifts than to buy them - and it costs less, too, but the best thing is that while you're making things, you're thinking about the particular person that particular present is for, and so good love gets sewn or hammered into each project."

Tom, Dick, and Mary said something that sounded like a snorted "Hmph" but could have been a thoughtful, "Hmm," and I said, a little lamely, that wrapping their own presents might do a little of the same thing. They said they weren't religious, anyway, so why should they do anything for Christmas?

"Because it's such a good excuse!" I said. "It's an excuse to give presents to people you love and an excuse to make good things to eat and visit family and be in touch with friends and have a good time. Good God, man," I said, losing patience. "It's an excuse to be joyful!"

Tom, Dick, and Mary Scrooge looked at me with apathetic astonishment. "Joyful?" they repeated. "Why?"

"Well, because," I began in exasperation, thinking I'd say, "because without an excuse for joy we might forget to include it," but then I gave up all the logical arguments because, after all, the answer was everywhere around them, anyway, if they would only see it.

"Because it's so much fun," I said at last. "Here, have some cookies. I made them just this morning."

Making Muumuus and Robes

I thought I had had a good idea for Christmas presents, but I worried as I sewed. Would my father ever wear a robe like this? Did my niece really want a muumuu for Christmas? Was I out of my mind to be making clothes out of Christmas fabrics? I thought maybe I had just gotten carried away by those beautiful materials in the fabrics store - bright gold angels on midnight blue, gold stars on green, green Christmas trees on red, red poinsettias outlined in gold, gold snowflakes on white. As soon as I chose one fabric, it was the next one I thought more beautiful, and so, there in the store, I kindled a plan: muumuus and robes for my entire family, sixteen garments in all. So I bought sixteen Christmas fabrics.

As I cut out the patterns, I worried about size. If Leah's muumuu was a medium, should Cameron's be large? If the robe pattern was unisex, did medium refer to a medium man or a medium woman? Was the difference between large and extra large a matter of girth or of height? I called my brother-in-law Billy in Georgia.

"Are you large or extra-large?" I asked.

"Extra-large," he said, and I went back to the store.

Worried about how much work I had before me, I went into a frenzy of sewing. I canceled all appointments, forewent all parties, and told a friend when she called I was sorry but I hadn't scheduled time for that phone call. Two days before I left for Georgia, my sewing machine broke down. I would have to finish my presents in Georgia, but I fretted about having to work when I wanted to be visiting. It was an unnecessary worry. My sister Sharon was still working on her presents, too, so we visited while we worked in her upstairs studio together, she at her desk and I at the sewing machine, back to back to preserve secrecy.

Appropriateness still worried me. "Nobody's ever going to wear a Christmas-print robe," I chided myself. But nothing could diminish my pleasure in working with those fabrics, and I began to think that what the gift was really about was providing a moment of Christmas delight. If we would all put on our Christmas-fabric clothes at the same time, I thought, if we would all stand together for just five minutes arrayed in this Christmas splendor, then I would be satisfied.

I wasn't asking that anyone like what I was giving. I was just asking for a moment of fun for all of us together.

When I reached the point of hand sewing, I wanted to sit downstairs by the fireplace, but I was worried about preserving secrecy. Would Billy ask what I was making?

"Not a chance!" Sharon laughed. "He wouldn't even notice."

So I sat in the rocker next to the open-hearth fire and across from the Christmas tree to do my sewing. Billy was on the couch, petting the dog; Sharon was upstairs making Christmas presents. We each had a cup of eggnog, and the Anonymous Four were singing Medieval carols on the CD player. I finished one hem, quickly shook out the robe and laid it aside, then picked up another. Billy said, "What are you making?"

"Sharon said you wouldn't ask," I said.

He looked sheepish. "I thought it was all right to ask," he said, "since it looked like some kind of robe or dress, and I knew it was nothing I would ever wear."

"Don't be too sure," I thought. Maybe for five minutes. That's all I was asking.

On Christmas morning my nephew Brian was the first to unwrap his robe. With an exclamation of glee, he jumped up and put it on. I gained hope. By then the others were opening their muumuus and robes, too. "How beautiful!" "Oh, it's a muumuu!" "Did you make all of these?"

"Yes, yes. Put them on. Everybody, put them on." In an instant sixteen people had slipped into muumuus and robes and bunched into a close group as though the fabrics themselves had pulled them together. We stood among tossed and crumpled Christmas wrappings in our Christmas patterns and colors like beautifully wrapped presents ourselves. Brian grabbed his camera and snapped a picture.

And that was it. That was my five minutes, and I was well satisfied. But I got a little more, too. Darryl said now he wouldn't have to walk to the bathroom in his house naked any more. Mom said she would certainly wear her muumuu when the weather got warmer, and Sharon told me that when Billy got home, he washed his robe, a pretty good indication, she said, that he would actually wear it. But it was my 25-year-old son, Ela, who pleased me the most. He wore his robe, green with bright gold poinsettias, all day long, like a smoking jacket. It was mighty handsome, and I was well satisfied.

A Sense of Other

Taking New Steps

A few weeks after his first birthday, my baby son took his first step. My baby! Holding on to the familiar chairs and tables, he started towards me; then, out of props, he hesitated, looked at me urging him on with smiles and encouragement, my eyes saying yes, do it, let go! Walk! He assessed the risk, the challenge, the reward, the distance – and he let go – he walked! All by himself, one whole step without any help of any kind, and then another, and his eyes lit up, and in that instant I saw that a whole new world had opened for him. Independence! "I can get there by myself!"

Sixteen years later and sixteen times more conscious of the meaning of that word "independence," that same boy took another giant step. Wheels go farther and faster than feet; motorized wheels propel feet the farthest fastest. For a year whenever we were in the car together, he was driving. He drove at night and in the rain; he drove on the freeway and in five-o'clock traffic in town. He drove many miles on rural roads. He stopped on hills and learned the clutch; he manipulated gear shifts, turn signals, dimmer lights, and windshield wipers. He took drivers' ed at school. He worked on week-ends and bought a car (my old car). By his sixteenth birthday, he was as prepared as any sixteen-year-old could be.

Letting go of chairs and table edges was nothing compared to taking a driver's license test. His appointment at the DMV was at 3:40. At 3:50 the heavens let loose with a downpour of rain. At 4:00 he pulled out of the DMV into late afternoon traffic and heavy rain for his road test. At 4:30, with a big grin on his face, he walked out of the DMV sliding his driver's license into his wallet.

When he took his first step, I picked him up and hugged him and put him in my lap and hugged him again and told him how proud I was of him. When he got his driver's license, I gave him a big hug and took him out to dinner. What a triumph! I had his car washed thoroughly for him inside and out, washed, scrubbed, waxed, shined, and odorized with a "new car" smell. After dinner he followed me home in his own shiny car.

And so Ela, who learned to walk on his own sixteen years ago, now learns to drive on his own. Independence! "I can get there by myself!" Given, of course, gas and insurance, oil, tires, and repairs, he can now go where he wants to if he wants to – whether I want him to or not. "No, you can't go" has got to have more force behind it now than "You can't go because I can't take you." If I can't take him, he can still get there. And if I have gained my own independence from those hours in the car driving him here and there, I have lost something, too. Those hours were shared moments of close companionship. Now, with his new independence, that place and that time for sharing is gone.

With a jolt I find myself making adjustments to a boy becoming a man with a man's independent means as well as a man's independent mind. Well, I never wanted to hinder the baby from becoming a boy; far be it from me to hinder the boy from becoming a man. There will come a time when this job of mothering will be over – not the relationship, but the job - and when Ela takes that third step towards independence, I hope he is as well prepared as he was for the first and the second. And I hope I can give him a hug, congratulate him warmly, and let him go, urging him, "Yes! Do it! Let go! Walk!" Because that's the way worlds open.

I Have No Enemies Bad Enough to Curse with the Flu

I looked like I had been ravaged on the streets of Hell. The flesh hung loosely over my bones; my cheeks were hollow; dark circles enunciated my sunken eyes. I had lost twelve pounds in four days, and I looked not slim but emaciated. My fingernails showed shockingly chalky white moons over pink skies. She who gazed at me from the mirror was but a skull-and-crossbones image of myself. Vaguely I wondered if I would have to bear this ugliness the rest of my life.

The flu was a pack of wolves which, chasing me silently, hounding my heels, had brought me down. Prostrate, I saw my ganglia tied in knots; I saw long, thin rectangles with sharp corners: boxes long-stemmed roses come in or the coffin one is buried in? I coughed and coughed, and from my childhood I heard my Aunt Zip say, "It's not the cough that carries you off but the coffin they carry you off in."

My 13-year-old son was a jewel. He kept the fire going and brought the wood in. He swept the floor, poured a bath for me daily and emptied it when I was done. While I was too sick to leave my bed, he emptied my pee pot. He made me broth and offered me oranges. He moved his bed downstairs to be near me if I needed help.

Then the wolves attacked Ela, too, and I moved my bed downstairs to be near him if he needed help. In our mutual misery we commiserated with each other, sympathized with and encouraged each other, and helped each other with spiritual as well as with physical healing. Too listless for games as strenuous as canasta or Boggle, lacking the energy to read for more than ten minutes at a time, we played ten thousand dice ten thousand times.

We drank water like winos on a binge, herbal teas and miso broth like horses at a trough. Sitting at intervals with the tent of a sheet over our heads, we inhaled the sharp steam of hot vapor baths that sliced through mucous in our noses and congestion in our lungs with the decisiveness of a surgeon's knife. Wrapped in sweaters and blankets, lying on makeshift beds close to the fire, we slept and slept and slept. The cat walked in and out, the water boiled on the stove, sun and clouds and stars in turn rolled across the sky, and Ela and I slept and drank tea and listlessly lay awake and slept again.

There came a time, before I was well, that we ran out of food and clean clothes. My insistence on daily baths and clean floors was ludicrous in the face of sweat-soaked sheets and dirty clothes, and if we didn't have some food in the house, Ela, at least, would starve to death. (I wasn't able to eat, anyway, but Ela never suffered from diminished appetite.) Reluctant to trouble anyone but in dire straits and real need, I called a friend. "I need help," I gasped weakly over the phone, and within an hour Joan was at my door with a hot dinner and a box of oranges and cabbages and, just as important, with smiles of concern and good cheer. She left with my dirty laundry and a shopping list and reappeared two days later with clean clothes for us and two boxes of fruits, vegetables, and juices. We were on the road to recovery. The wolves were skulking in retreat; my ganglia were making smooth connections again, and visions of long rectangular boxes had faded.

With a bit of strength repossessed, I could read aloud again and so turned us to the literary healers. Each episode of *My Family and Other Animals* laughed us one step further along the road of recovery. By the time we had finished the book, we were jubilant in victory. Like hot broth through the digestive tract, laughter had poured life into our spirits. Vitality coursed through our veins again. We were giants in health. As a congratulatory reward, we treated ourselves to a night at the movies – *The Gods Must Be Crazy,* which was the conclusive routing of the wolves of the flu.

Working with Men

Working with Jerry (now an ex-boy friend) was an exercise in maintaining self-worth in the face of a supercilious, sneering attitude. He would give me contradictory, sarcastic directions until I couldn't tell what he expected of me. Was it "Don't be stupid. If you would just help me, we could get the roof on" or "Don't be stupid. Of course, we can't put the roof on by ourselves. Go ahead. Climb the ladder. When you fall off, you'll see I was right"? Whatever move I made would be wrong. When I told Jerry just to talk straight, without sarcasm, I got the mock-soft, condescending, overly patient voice one uses when dealing with a difficult child, giving meticulous, unnecessarily detailed directions slid towards me on slivers of sarcasm.

Working with my son's father had been a different experience. Dan was a talker. A building project for him was a chance to relate, to work out ideas by talking them through, to discuss the abstract wonders of philosophy or education. He was a good builder, and we worked well together: I worked and half listened while he talked and half worked.

It was after Ela had grown up and moved to Seattle that I decided to put new interior walls in my house. When he heard what I wanted to do, he managed to take a few days to come down to help me. Working with Ela was neither like working with Jerry nor like working with Dan. It was like working with my father. Like his grandfather, Ela is quiet when he works, absorbed in the work itself. It is not an exclusive silence; as with my father, I felt he was not withdrawn from but present with me. Like my father, he would answer questions, respond to short snippets of conversation, laugh at ironies. But mostly he was intent on the work.

Ela is far more skilled with tools and has a better understanding of construction than I, so he was the main builder and I the assistant. But he wasn't a very good foreman, and I had to discover my own jobs - holding one end of the board while he put up the other, hammering boards in place while he measured and sawed, measuring and marking boards for him to saw and put up. If he was busy with a project that didn't have a place for me, I turned elsewhere - making a sill for a window, preparing the next wall, keeping the work area safely free of clutter.

I was impressed with the careful way Ela worked and with his patience - again, like his grandfather, who is renowned for his patience. When a board didn't fit, Ela, unflinchingly good-humored, cut it again and if necessary again and again. He was considerate of my wishes ("Do you like this board here? Should I do it like this?"), made creative suggestions ("If we take down this plywood walkway, light will fall through the skylights to the floor again"), and offered artistic solutions to problems ("We can use graceful, twisty, dark-red manzanita branches for shelf supports"). He was never irritated by me - not by my indecisiveness or lack of skill or insecurities in thinking he was thinking I was being stupid. But of course, Ela wouldn't be thinking that. That was Jerry. For Ela, I was an equal partner in the project, not because I am equally skilled but because I was doing, for my capacity, equally well. Working with Ela was an exercise in working well.

"This Is a Bowl My Father and I Made"

After I helped my 89-year-old father lug a 90-pound round of cherry into his shop, he picked up a smaller cherry log and turned it over in his hands. "I think we could make a bowl out of this," he said.

"We could?" I said carefully so as not to disturb the pronoun, which was generous.

To begin, my father cut off one end of the log "square to the future center line," as he put it. He considered the wood again, then nodded. "I think there might be something interesting in there," he said, meaning, I think, an interesting grain. I was worried about a crack that ran into the wood, but my father hadn't mentioned it, so I didn't, either.

He mounted the piece on a faceplate, which he screwed onto what he told me was the lathe spindle. Then he picked up a scrap of paper and a stray pencil. "How tall do you want your bowl?" he asked. "What kind of stem do you want on it? How wide do you want the mouth? How big the base?" He made some sketches and adjusted them to my directions, then considered them, nodded, and turned back to the lathe, whistling a bit of Beethoven.

The function of a lathe, I saw now, is merely to turn the wood while the woodworker works the hand tools, which my father named for me as he turned the outside shape of the bowl: skew, gauge, scraping chisels, cut-off chisels. My father held these tools steady against the spinning log. Sawdust flew as a roundness began to form.

Next my father put the bowl on his drill press and drilled a hole into the inside, then remounted the bowl on the lathe and took up a new tool to cut the inside.

"I made this tool myself," he told me as he set the lathe spinning again. "I drilled a hole across one end of this three-foot piece of three-quarter-inch iron pipe. That was for this metal turning bit, which I secured with a set screw. Then I attached a steel bar at the other end to resist torque forces." The crossbar rested on a platform

93

attached to the lathe bed. With this tool my father could reach deep inside the bowl as it turned, taking shape, spitting sawdust.

Since my father had used a generous pronoun, I thought of asking to try my hand at the lathe, but watching my father's precision and delicate mastery of his tools, I thought how easy it would be to thrust a chisel through the side of the bowl and decided not to ask. I asked instead how he learned woodworking.

"In shop at school," he said. "When I was eleven years old, I made a porch swing for my family. They sat in that swing for years." Almost eighty years of experience was turning our bowl.

My father stopped the lathe and ran his hand around the inside of the bowl, then gave it to me. It was smooth and even. The walls were gently curved; the lip sloped delicately inward; the bowl rose gracefully to a six-and-a-half-inch diameter from its smaller base and short stem. The chisels and skew had worked through the crack into solid wood. There were interesting circular patterns in the wood, variations of color from cherry-wood red to stripes of dark brown, splotches of amber, and lines of black.

"It has to sit in a solution for a few months, then be sanded and shellacked," my father said. I believed him; what did I know? But a week later, on Christmas morning, he handed me my bowl, smoothly sanded and well finished.

I have other bowls and goblets my father has made, but none is so fine as this one. I proudly show it to friends. "This is a bowl my father and I made," I say each time, each time knowing I had done nothing. Nonetheless, I'll use my father's generous pronoun. This is a bowl we made.

A Portrait of My Mother

She didn't like to cook, yet her family ate not TV dinners but beef stew, fried okra, and peach cobblers. She didn't like to sew, but her four daughters wore stylish handmade dresses, and every Easter four stairstep little girls modeled new handsewn Easter outfits to church. She didn't like to drive, but day after day, week after week, year after year, she drove to school or church to pick up her children, two at a time, one at a time, from band practice, Girl Scouts, church activities, parties, club meetings, sports. She didn't like to clean house, but her house was always passably clean and periodically neat, since she ran the sort of household five children were free to be children in.

Perhaps that is anybody's mother, though, because isn't that what mothers do - sweep the floor, wash the dishes, clean the clothes? That was not what made people say to me when I was a child, "I don't see how your mother does it." What they couldn't see how she could do was raise five children and paint, too, because what my mother really liked to do was paint.

With a love for old things, an artistic bent, and a creative curiosity, she was always making something useful or ornamental out of nothing much. By adding this and that, by painting designs of grapes or flowers, she turned chicken feeders into candle holders, milk cans into lamps, tin cups into pencil holders. Haunting the Sandy Springs junk shops, she hauled home old trunks, tin boxes, chairs, milk cans, school desks, shelves, bureaus, anything of one man's junk she could transform into another woman's treasure. Asking my father if he would put a hinge on the trunk and fix the rungs of the old chair (he would), she would paint it and antique it, then sell for a handsome figure what she had bought for a song. The house filled with painted junk turned treasure, and my mother became well established as one of Atlanta's foremost tole painters.

That was the mother I was proud of, but the mother I loved was a more intimate and personal woman. It was she who, immediately after her annual open house for selling her work, would give each child a check from her profits. "Here is some foolish money," she would say. "Don't spend it on anything sensible." She slipped I-love-you notes into my sack lunch for school, gave me a warm hug around the shoulders as we left the car to go shopping, brought me tiny surprises from town, hung a "welcome home" sign on the door of my room when I came home from college. The best of my mother, for me, was that it was she who taught me how to say, "I love you."

When I was growing up, I had a special love for the back-yard pear tree blossoming whipped-cream white in early spring. When I was away at college, each February my mother would send me in the mail a bundle of pear branches. I would put them in water to force the blossoms, spring before spring was meant to be, a bit of home in Tennessee, a bit of my mother saying, "I love you." That's what I mean. No one else has taught me so well how to say, "I love you." And no one has ever taught me anything more important.

Gathering the Rainbow in Pots

As I was walking through the charming little Danish town of Faaborg with my Danish friend Maren, I peeked through a brick archway into a courtyard tumbling with flowerpots and drooping with greenery and bright flowers. It was Maren who led me in.

Through the courtyard was a tiny pottery shop. Scores of flower pots in bright colors crowded the shelves on the four walls of the square little room, the red ones here, the yellow ones there, the blue ones vibrating like the sea on the center table. The brilliance of these nonearthy hues with their subtle shifts of color contrasted with the earthiness of the clay. Like the hats of the Cat in the Hat, tall, slender flower pots were stacked one into the other, surrounded by rippling stacks of squat pots and piles of little pots and big, rotund pots; they overflowed the table and shelves to lie in riots of separated colors under the table and in the aisles. Stuffed between and around the pots were twisted unpolished metal sculptures, starfish dyed the same bright colors as the pots, unusual seashells, and a pot full of dried seahorses on sticks.

"I love these pots!" I said.

"But, of course, you must buy one, then," Maren told me.

"Oh, no," I said, thinking of my limited funds. "I shouldn't, really."

"Isn't this beautiful?" Maren said, holding up a yellow pot that looked like sunshine on wheat. "I think you should buy it."

"Oh, but that would be foolish," I said. "I'm flying home. Pottery is too heavy to carry in a suitcase."

"But you can take an extra bag," Maren said, setting down the yellow pot and picking up a red one.

"I really do love these pots," I said again, admiring a blue one.

"You really should buy one," Maren urged.

There was something Maren knew that was more important than lack of money or difficulty of travel. I wasn't sure what it was, but, overcome by the beauty of the colors in the flower pots flying like butterflies around the tiny, crowded room, I leapt into faith.

97

"Okay, yes," I said. "I will. I'll buy this pretty little red flower pot for Leah and Ela. It's the perfect red for their room."

"Yes, but you must buy something for yourself," Maren said "Don't you just love this yellow?"

"Oh, yes! Okay. I'll buy one of these yellow pots, too," and while I was choosing which variation of the yellow flower pots I wanted, Maren placed a blue pot in my other hand.

"And this one, too," she said. "You really did like this one; you should buy it, too."

By this time I was giddy with my purchases. "Yes!" I said. "I'll buy the red one and the yellow one and the blue one," and it's a wonder I didn't say "and an orange one and a green one and a purple one, too!" but my arms were full, anyway; I couldn't have held another pot. I stood in line at the counter in the crowded little shop, my arms full of flower pots, my face glowing.

"Look at this!" Maren said, standing in line behind me. "These candles match the pots. You really must buy some," and she placed a pair of yellow marbleized candles in my already laden arms.

"Then give me a pair of red candles to match Ela and Leah's pot, too," I said. I had no idea how much this was going to cost, but purchasing was out of control now.

There was some confusion about the money rate, since all I had were travelers' checks and the man at the counter wasn't used to changing money, and in the end I'm not sure how much I actually spent, and Maren threw in I don't know how much of her own money to make it come out right, but when I walked out of the potter's shop, I was doing little dance steps. I was so pleased with my purchases! And I never would have bought anything in that little shop that I liked so much if it hadn't been for Maren.

The following week Maren, her husband, Lasse, and I spent an idyllic day on an island off the west coast of Sweden, where the sea was Grecian blue, white sailboats sailed on the breath of angels, and the picnic on the rocks tasted like fresh air and mountain springs. At the end of the day we went into a painter's studio, and there Maren was as I had been in the potter's shop. We all studied the paintings, choosing as favorites first this one and then that one, and finally Maren chose to buy a painting of a spot near where we had had our picnic, with the white rocks and the blue sea and the white sailboats. As we walked out of the shop, Maren, carrying her new painting, said, "Now I will always be reminded of an idyllic day on Marstrand."

So I know, now, what it was that Maren knew about buying things that I hadn't known. When I bought my pots, I bought more than beautiful flower pots with matching candles, and these pots hold more than the plants I've put in them. The whole of Faaborg is in those pots: the narrow streets and brightly colored houses, the thatched roofs and hollyhocks, the inn where we stayed, the potter's shop where colors danced like butterflies, and Maren.

Reparations for a Grievous Error

I had made a terrible mistake. I had accused a friend, my firewood vendor, of giving me green wood. He was deeply insulted. How could I have thought he would do such a thing?

I tried to explain: it was the middle of the night, it was raining, I had gone out in my nightgown and rain poncho to cover the wood with a tarp. I had had only a flashlight for light, and as I tried to push all the wood under the tarp, I had thought, "This wood smells green. And it's heavy, like green wood."

Burt said he knew it wasn't green; he had carried it himself, so he knew it wasn't heavy.

Burt is one of the strongest men I know. What he calls heavy and what I call heavy are not the same item.

I told him I had thought maybe he had brought me the only wood he could find, knowing how desperate I was to get firewood up here before my road became impassable in the rain.

"I sweated rocks getting up that hill," he said.

I told him how grateful I was and said he always brought me good, dry wood and that I always spent every winter grateful every day for his wood and that no one else would ever be so good to me, and finally I just gave him a hug and said, "I hope things are clear between us now."

He said, "Yeah, I guess," and I knew they weren't.

Three days later, October 31, I spent the entire day making cookies: brandy-laced joggers, fudge-filled butter cookies, date-and-orange bars, and marshmallow fudge brownies. I piled dozens of cookies high on a platter, which I covered with tinfoil and tied with a blue ribbon.

Then I reached up to the wall behind my desk and took down a mask, a gold-sparkly woman's face with a large bush of black hair, black lips, black eyebrows, and black-lined eyes. I put on red shoes, a black leather mini-skirt, a gold lamé blouse, and a red feather boa, then slung a grey cloak over my shoulders and left the house with my mask and cookies.

It was raining hard when I pulled into Burt's driveway. He lives by himself in a little house on a lonely country road in a spot as remote as my own. I got out of the car, put on my mask, and picked up my cookies, then walked across the yard and knocked on the door.

Burt opened the door, and the mysterious woman walked in.

Burt said, "What the hell?"

"Mr. Burt Hayles?" asked the mysterious woman in a strange, deep voice. Burt hardly owned up to his identity, but she went on, anyway. "A certain Diana Coogle has asked me to present this to you. She is deeply sorry for the mistake she made and asks your forgiveness. She would have come herself, but it is raining out, and she is sitting warm by her fire, which is burning brightly with good, dry wood. At any rate, she most humbly begs your forgiveness. She is, you know, prone to hysterics and sometimes jumps to precipitous conclusions. She has so much appreciated your excellent service in the past and hopes she can call on you in the future. Please. Accept this gift as her apology. She begs you."

Burt didn't say anything, but he was smiling. He accepted the cookies. The mysterious gold-faced lady left the house to return to her who had sent her, with the message that both the error and the insult were most likely forgiven.

Left-over Lover

I hate Valentine's Day. All the little red hearts and cutesy, mushy messages of schmucky, moon-eyed lovers, the flower vendors selling red roses on the sidewalk - "Send your sweetheart a dozen red roses." I want to send him a dozen dead roses.

Last year he gave me flowers and a card: "To the girl I love." For a year the card was pinned to my bulletin board. Who gets the flowers this year? Who, now, is the girl he loves? Who is the lucky Valentine of the Year?

What difference does it make when it's meaningless? Short-lived love is a loser's game.

I should have known. When his Christmas present arrived shattered in the mail, I should have known. Maybe he even broke it before he wrapped it in bright, joyful Christmas paper and sent it to me deliberately, symbolically. Only, he didn't say that. There was a lot he didn't say, a lot I didn't know.

I guess I should have known, too, when the bookcase fell over and all the books fell out and the glass box and the delicate glass jar of rose petals from my parents' forty-seventh wedding anniversary and the plant he brought with him when he moved in with me - all flung out of the shelves onto the floor, and the only thing that broke was the plant. A beautiful, full, many-stemmed bit of greenery, branches now broken, its long, trailing, sensuous arms dangling at cracked joints, and even still I pick dying leaves off its thin, diminished beauty. I should have known. I ought to take things symbolically.

I hate the songs that say, "Karla, we can make it if we try." It isn't true. I had thought it was true, and I did try, and we didn't make it, but I guess we can't make it unless we both try.

And I hate the songs that say, "He's gone from me, oh-oh-, tragedy," because they're not true, either. They belittle a thing by making it bigger than it is. It isn't tragedy. I won't let it be destructive to me, my life, my philosophy, my emotional stability. I won't. It's not worth it.

I can make it if I try.

I'm on a pendulum of responses, vacillating between slashes of the sword. One slash, and I want to tell him what a creep and a coward he is, what a cruel, uncaring, insensitive cad, how petty and thin-visioned - so easy to forget me when I'm not there for his immediate gratification, and the other slash: No, sir! He's not worth another tear, not another moment of misery, and he'll never have the ego-gratification of thinking I care a fillip whether he loves me or not. He'll never bask in the illusion that I can't get along without him. I can.

Yet I can wallow in the self-righteous ego-satisfaction of knowing I was right and he was wrong, that while I was good, kind, loving, and loyal, he lied, skulked, and pretended. So what happened? He, the wretch, is blithely happy; I, the good, suffer.

There was his picture to deal with, too, hanging on my clipboard over all his letters. I couldn't quite throw it in the fire, and I tried to turn it to the wall and couldn't, but I did turn last year's Valentine's card ("To the girl I love") face down and smash it over his face, a big square blank now with a mustache drooping under it. And under the mustache, on the exposed jugular vein where his unbuttoned shirt reveals a tantalizing touch of sexy chest, twangs a dart.

Some day I'll exhume his buried image and the love he so carelessly threw away and look at them and smile and feel warm again. But not today. Maybe next year red hearts and flowers will bloom with smiles, and maybe the idea of spring will lighten my winter-tired tread again, but now, today, this year, February is cold and Valentines are pierced with icicles.

The Guru in the Bathtub

In 1969 I was living with my common-law husband in a commune on Skyline Boulevard in the Santa Cruz Mountains, two miles down the road from La Honda, California. Ken Kesey's name hung in the ocean-misted air not only because we lived so close to the famous home of the Merry Pranksters, not only because *One Flew over the Cuckoo's Nest*, like *Autobiography of a Yogi* and the Don Juan books, was required reading for a hippie certificate, but because Ken Kesey was a pioneer of the hippie culture, legitimizing with his literary credentials and seriousness of quest our experiments with psychedelics, body paints, long hair, outlandish clothes, light shows, strobe lights, eastern mysticism, wild use of color and patterns, full-moon festivals, solstice celebrations, free love, and back-to-the-land living. We were high with the breakthrough freedom we had found and giddy with our own importance because, although the hippie movement attracted some riffraff and ne'er-do-wells, it also rested on a foundation built by intelligent, spiritual truth-seekers, chief among whom was Ken Kesey. He was our guru.

He was more than a guru of image, though; he was a guru of language as well. He gave us lingo. "You're either on the bus or off the bus," we said to each other, referring to more than the psychedelic bus that took the Merry Pranksters across the country, and "cuckoo's nest" was a common phrase, implying, as it did, the question of which side was crazy, anyway. Kesey's down-home vernacular, the frontier values of his characters (self-reliance, courage), and his folktale language of American storytelling validated our sense of being rooted in an American tradition and, more specifically, in a tradition of the American West. The sense of quest Kesey depicted in his novels was our underlying *raison d'être* and gave us the courage to believe in ourselves.

By 1972 Dan and I had left the commune in the Santa Cruz Mountains and were living here in the Siskiyou Mountains of southern Oregon. We knew, of course, that Ken and Faye Kesey and their kids were now living on a farm near Springfield, Oregon – close

Springfield, Oregon – close enough that we decided one week-end to visit them. Although *Contemporary Authors* claims the Keseys were "plagued by visitors to the farm, sometimes several hundred in a week-end," we didn't know we were part of a plague and knocked excitedly on the door. When we asked for Ken Kesey, we were told, "Oh, yes. He's taking a bath. First door on the left. Go on in." So we went in, and there was Ken Kesey in the bathtub, spinning philosophy peppered with Keseyian folk language for three or four (not several hundred) visitors.

Let me therefore now dispel, once and for all, the popular image of the dirty hippie, for I have seen the guru himself in the bathtub.

Though I am appalled now to have been a part of such a large pilgrimage that the Revered himself couldn't even have a bath in private, Kesey holding court from the bathtub is an unforgettable picture from those days that I'm not at all sorry to have stored in my memory. Besides its drollery, it shows both how unconcerned Kesey was about conventional boundaries and how generous he was in treating people with more respect and welcome than they probably deserved.

After the bath, Kesey dried himself off and got dressed; then everyone at the farm - Faye, kids, visitors, friends, neighbors, family – followed Kesey out to the pasture for a neighborhood, Sunday-afternoon softball game in the pasture. It may not be conventional American practice to visit in the bathroom, but there is nothing more apple-pie American than sandlot baseball. In a way, that game in the field with all the kids proved Kesey more rooted in American values than defiant of those values, and so, too, in its heart, was the hippie movement.

Like Good Nazis

Early spring brings an itch to the feet. Some of us just want to scratch that itch with a shovel; others want the foot on the road. I'm really more of the former type than the latter, but one spring I was on the road, anyway, on a Greyhound bus coming home from Portland.

The bus was crowded and the bus driver snarly. He reminded us there would be no foul language or smoking. After an hour on the road, he spoke through the microphone again to tell the people in the back of the bus to cut down on the noise. I hadn't thought they were disruptive.

We had a brief rest stop in Eugene, and when we reboarded, the driver told us Greyhound was sending out a second bus, so there wouldn't be any new passengers on this one. But when he counted us, we were one too many. Griping and complaining, he walked through the bus checking our tickets. He found the culprit, a confused Vietnamese man with several plastic bags and a duffel bag. The bus driver told him to get off; the man, frightened, clutched his duffel bag to his chest and said something in an incomprehensible language. "Does anybody speak this man's language?" the driver barked at the rest of us. We all sat there in silence. The driver shoved at the man to move him towards the door. The man, shaking in fright and bewilderment, scrambled off the bus.

"He left his bags," said a very tall white man in the seat in front of me.

"Those are his bags," someone else said, but the driver either did not hear or did not care. He climbed querulously into his seat and started the bus.

A wave of concern flowed through the passengers. The poor Vietnamese man who was shoved out the door - he's left his bags on this bus. But the bus was on the road, headed for Heaven on Earth, the restaurant where we would stop for dinner.

Why hadn't anyone immediately grabbed the bags and jumped off the bus to give them to the man? Why had we all just sat there? The bus driver wouldn't have left us. What kept us from act-ing? I was appalled that no one had moved to help. I was especially appalled that I hadn't done anything.

Maybe we were behaving out of a misguided sense of doing right. We were supposed to be in our seats, ready to go. The bus driver didn't like things to go wrong. He wouldn't like keeping the bus waiting while one of us took the bags to the Vietnamese man. So, like obedient people, like docile sheep, like good Nazis, we stayed in our places.

In a 1988 essay Mary Lee Settle tells of being at a small dinner party in London immediately after World War II. One of the guests, an elegant French officer, taking it for granted that he was making an acceptable remark, said, "Well, at least Hitler did one thing for us. He got rid of the Jews in Europe."

"I was too frozen with shock to move or speak," Settle says. "I felt drained of life. Despair can leave you too lost to resist seduction. You go along. I did not leave quickly enough. In short, I was polite."

I was polite, too, on the Greyhound bus, but to the wrong man. Just as no one at the dinner party dared to confront the French officer with his enormous assumption, no one on the bus dared to help the man in need in the face of the bullying man in charge.

When we got to Heaven on Earth, the tall man in front of me carried the Vietnamese man's bags to him on the bus behind ours. That the Vietnamese man got his things back makes the story turn out right in the end, but it doesn't alleviate the guilt of all of us for not having done right in the first place. If we can't step forward in the name of justice in small matters, for the love of God, what will make us do so in matters of greater consequence?

Giving A's

Those A's feel so good. They make your heart glow and put a smile on your face. They give you a sense of accomplishment and worth. Students like A's, and I like giving A's. I should give everybody A's all the time, and everyone would be happy.

But, of course, I don't. I try instead to use a grade that honestly reflects the quality of the work, and I write detailed comments in the margins of papers to help students improve their writing. But many students expect A's, so handing back the first papers of term takes psychological armor on my part. "A D!" one woman exclaimed. "I never made a D in my life!" and she stomped out of class and never came back. Another woman, with more grace and more wisdom, looked at her grade and gulped. "I have a lot of work to do," she said and dug in for the duration, applied herself diligently, and was writing A papers by the end of term. One man told me he had been manager of a big business for two decades, so how dare I not give him A's? The most astute comment came from a young man who looked at his grade and said, "Oh. It's not high school any more."

But if low grades are marginally acceptable on first papers, I had better do better with subsequent grades. Students explode with exasperation at my slowness to understand what an A paper is. "Do you know how much time I spent on this paper?" they demand angrily. Oh, of course. How dumb of me. Effort equals A. I had thought effort equaled A if and only if effort equaled quality. But I am dimwitted. Forgive me. Yes, it's still a C paper.

In one particular composition class my students were dangerously sulky and restless the day I returned their papers. Class participation was shot through with pugnacious reluctance. Student belligerence was gelling into student rebellion. The buzz was unvoiced, but I could hear it: "We demand our A's! We have a right to A's! Give us our A's! Give us our A's!"

At home, grading their new essays, I got angry, too. They wanted A's? All right, I'd give them A's. I read each paper and put an A on it and not another mark - neither correction nor comment, neither criticism nor praise. They wanted A's, and that's what they got.

The increasing buzz of amazement as I walked around the room handing out papers the next day was even more audible than the bellicosity had been. "What'd you get? An A? Good for you! You got an A, too? So did I. And you did, too?..." By the time I had returned to the front of the room, they were looking at me accusingly. "We all got A's, didn't we?" they asked. They didn't sound happy.

"That's right," I said. "You wanted A's, so you got A's. You can keep the A if you want to, or you can give your paper back to me for an honest evaluation and an honest grade." To my surprise two students immediately jumped up and put their papers on my desk. By the end of the class all but two students had returned their papers. (Those two young men decided they needed those A's.) But my experiment had proven something to me. Although students want to feel good by getting A's, most are more interested in learning what they have come to class for and prefer their A's, when they get them, to be honest.

Wedding Quilt

One of the many things in my life I am grateful for is that I have remained friends with Ela's father, Dan, in spite of our separation when Ela was two and that his present wife, Tracy, has also become my friend. Such relationships helped us all be better parents while Ela was growing up and disallowed any reticence for Tracy to suggest to me, when Ela told us he and Leah were getting married, that she and I coordinate making a quilt for them. I immediately concurred. Together we would shop for fabrics, which she would pay for. We would draw up a list of family and friends, each of whom we would ask to contribute a square out of the fabrics we would send them. To mitigate the anticipated wails of protest from those who did not sew, I would suggest they could paint a square or use a stencil or fabric markers or do anything else creative. When the finished squares had come back to us, Tracy and I would sew them into a quilt.

Everyone we contacted said, "What a great idea; I'd love to make a square."

Our friend Louann joined Tracy and me as Quilt Coordinators. We designed the quilt in a checkerboard pattern of green and maroon (the wedding colors) with eight rows of six squares each. For the center we would use four squares in a wheel of fortune pattern; two blank squares would represent Leah's father and Ela's stepsister, both recently deceased. The remaining forty-two squares would come from the friends and relatives to whom I sent the necessary fabrics along with what I thought were precise instructions and the deadline date underlined twice, printed in bold, and mentioned three times.

Soon the squares started coming in: a gold-outlined appliqued tree; an embroidered Mayan "Hunab Ku, principle of life beyond the sun"; two entwined hearts; a Burmese prayer-rug square. One friend made a four-petaled flower using fabric from her "old hippie wedding dress." My mother painted one of Ela's sculpture-instruments on her square; my sister Laura, author of a book on the folklore of flowers, depicted ivy, the symbol of wedded bliss, on hers.

Other squares were more cumbersome. One held two beautiful handmade dolls, replicas of Ela and Leah. Another had round bits of cedar sewn into a wreath shape. Another was "Newlyweds' insurance": a penny for money, a little bag of rosemary for luck, a piece of wood for shelter. Oh, dear, oh, dear, oh, dear, I thought. This is a *quilt*.

I had to do some last-minute urging, but in the end all the squares arrived, and Tracy, Louann, and I spread them on Tracy's floor to determine their order. Flat ones had to balance three-dimensional ones, bulky ones to go at the bottom, the bright whites and golds not to be bunched together. A beautiful square of a doubly-entwined calligraphy of "Ela and Leah" in bright, thick gold paint from my sister Sharon, a professional calligrapher, went in the center of the wheel of fortune.

Then I started sewing. I fixed the squares that didn't conform to directions, resewing some onto plain fabric and resetting margins on those that had none. I made darts in the squares that were puckered to make them lie flat. Louann cut strips of fabric for the Seminole patchwork borders, and I sewed and sewed and sewed. As I sewed I developed a relationship with each square. I loved the flat, easy ones but grew to appreciate the bulky ones, too, as whiffs of sweet cedar from the wooden wreath or rosemary from the Newlyweds' Insurance drifted by or the bells on the "jester's star" tinkled faintly.

Two weeks before the wedding, Tracy, Louann, and I sat on my lawn in sunlight filtered through cherry blossoms and sewed and sewed. The breeze stirred the wind chimes into song; the cats sat on the roof above us; the daffodils and grape hyacinths splotched the woods below us with color. As our fingers flew, we talked about Louann's daughter's budding romance; about Tracy's daughter, who had recently died; about Ela and Leah and the upcoming wedding. We had been working together since morning; now we could feel dark approaching. Without electric lights, it was essential that we finish before dark. Wielding my needle faster and faster, I felt like the god in the myth who held back the sun till the task had been accomplished. Just before dark, we stitched the last bits, shook out the quilt, and spread it one last time on the grass.

It is the ultimately beautiful quilt, a sheath of blessings for the bride and groom from forty-three people who love them dearly - and maybe especially from the mother and stepmother of the groom, whose joint project it had been.

A Traditional Thanksgiving at My House

Years ago I built this small house on the mountainside. Life is simple here, by preference, but when Ela called from Seattle to say he and Leah would be home for Thanksgiving, I thought, "This year I want to have Thanksgiving at my house." But could I have a traditional Thanksgiving in such a house?

Traditionally, Thanksgiving means family. If my son is my family, he is equally my ex's family, and if my ex is married, his wife is family, too, so family for Thanksgiving would be Ela and Leah, Dan and Tracy, and I. Shel, a family friend, would also join us.

Traditionally, the family eats at a table, but with only a fold-up card table, would I have to limit Thanksgiving dinner to sweet potatoes, soup, and bread, buffet style? "Cook those pies!" Ela said. "Don't worry about a table," and he disappeared outside.

Traditionally, those who cook, cook in a kitchen. Here, I sat on the built-in couch under the south windows rolling out pie dough on the trunk while Leah, opposite me, prepared sweet potatoes at the sink. Wild rice bubbled on the stove under the bedroom loft, and when I took the apple-pecan upside-down pie out of the oven to put in the cranberry pie, buttery, spicy steam lifted towards my bed. With the sweet potatoes waiting for oven space, Leah started on the green beans. I pureed pumpkin through the strainer my sister had sent me as part of her "Queens-in-art" gift for the cook without electricity, then set the pumpkin mousse, creamy smooth and orange, outside the back door, where the cold autumn air served as a refrigerator.

Traditionally, one sits in chairs at Thanksgiving dinner, so when Ela hauled in his table with its old wooden-ladder legs, I looked around for chairs. My sit-on-your-knees desk chair was one, the ordinary chair at my sewing machine another, and the bench from the front deck a third. We could use the milk can I kept kindling in if we put a pillow on it, and Ela said he would sit in the little rocking chair, but we were still one short. "Don't worry; keep cooking," Ela said.

Traditional Thanksgiving foods began splashing the house with colors and smells: caesared green bean salad, sharp with sherry vinegar; Burgundian holiday bread, spicy with fennel and ginger; deep yellow, sherried sweet potatoes; wild rice salad dotted with dried cherries; dark red cranberry pie with its white sour-cream topping. Ela came in with the sixth chair: a stool made from some curved iron pieces and a round of wood. We spread the table with my grandmother's white lace tablecloth and set it with amber glass plates; green napkins in deep pink, silk-rose napkin rings; and purple, blue, and red candles. We put extra candles and kerosene lamps around the house. Ela used my Swiss Army knife corkscrew to open the Merlot and let it breathe.

Traditionally, there is a turkey. I'm vegetarian, but my guests were bringing a turkey stuffed with rosemary dressing, and when I opened the door to Tracy, Dan, and Shel, those luscious, warm smells rushed in with them. Finally, with the turkey and dressing reigning at the head of the table, salads, vegetables, breads and relishes sashaying between the rows of plates, and the room sparkling with candlelight, the family squeezed around the table, smiled at each other, raised voices in thanks, and sat down to eat.

Traditionally, one looks at one's family and the bounty spread on the table and feels abundantly thankful. This was, wholeheartedly, a traditional Thanksgiving dinner at my house.

A Sense of Self

Mitochondrial Eve

This winter, visiting my mother in her home, I was doing some hand sewing while we talked. Suddenly she said, dismayed, "But you're sewing backwards! You're holding the material upside-down and should be sewing left to right, not right to left."

It didn't seem backwards to me, since I've been sewing that way for decades, but it does seem odd that I would have reversed the direction for sewing my mother had taught me. When I thought about it, though, I found other things I do backwards. Teaching class, for instance, I'll start writing on the far side of the blackboard, then move to the left, writing Chinese-like, right to left. Am I responding to genes bequeathed me by an ancient Chinese ancestor nowhere apparent in my genealogical chart?

When I was a student at Cambridge, I roomed with a Japanese girl named Yoko Gibo, who had a wardrobe of beautiful kimonos in her closet. One day she dressed me in the kimono of my choice, which was decorated with flowing birds and brilliant flowers, and she added all the proper underclothes, shoes, and cummerbunds, too, and pinned my hair up to show off the nape of the neck, as a Japanese girl would do. She taught me to walk in the Japanese manner with little, slipping steps that suggested demureness and deference without loss of dignity. Looking like sisters in our kimonos, we walked through Cambridge to Queens College for a Japanese tea ceremony. Yoko was surprised at how easily I had acquired the Japanese manner. As for me, I felt I had come home to my Japanese self.

My short stature helps me look Asian, but I have been told, especially years ago when I wore my dark hair in braids, that I look Native American, too. I would like to think that my understanding of the natural world as a spiritual home is further proof of a Native American ancestor, but search as I might, I can't find that great-great-grandmother. And if my stature is Japanese and my features Native American, my gestures are Italian. I use my hands when I talk, vividly and broadly, like a character in an Italian film, though as far as I know, there isn't a smidgen of Italian in me.

Did I in some previous life live in Italy? Was I a geisha in Japan, or did I in some other century marry a Chinese potter? I have never adhered to theories of reincarnation or the transmigration of souls, but the strength of my experiences has always seemed to demand an explanation.

Last week, to my delight, I read about Mitochondrial Eve in Stephen Pinker's book, *How the Mind Works.* Although Pinker admits the claim is controversial, he also contends that "evidence is mounting" to indicate that the mitochondrial DNA (that which is inherited only from one's mother) of everyone on the planet can be traced back to one African woman, dubbed Mitochondrial Eve, who lived between 200,000 and 100,000 years ago and who, in the all-maternal line, is the ancestor of us all.

Can it be so? If it is, then the mystics are right! We are one! We are all brothers and sisters! I am related to the Chinese and the Italians, the Japanese and the Native Americans. I don't have to sit on my hands when I talk or relearn how to sew. In the manifestation of my personality, I am merely reflecting my ancestral mother, the Mitochondrial Eve - and so, whether you know it or not, do you.

Calamity with Clocks and Calendars

It's easy to see how I made the foolish mistake I made this past week, if one considers the human fallacy element in using clocks and calendars. On Monday, after a day in Grants Pass, I gave an orientation at 6:00 at Rogue Community College for the telecourse I'm teaching this summer. Everything went well, except I didn't have a chance to do my laundry, but I thought I could do it in Medford the next day before giving the telecourse orientation at RCC there. If orientation began at 6:00 and I got to Medford at 3:30, I would have about two hours for laundry and plenty of time afterwards to find the building where my class was scheduled.

At 5:30 I finished my laundry, well pleased with my timing, and then looked at my calendar, where I had written down which building my class was in. That's when I discovered my ghastly mistake. This orientation wasn't between 6:00 and 7:00 like the one in Grants Pass; this one was between 5:00 and 6:00. Oh, what had I done! "How stupid, how careless, how irresponsible," I berated myself as I dashed to the college in case some students were still hanging around. But the lights were out in my room, and, in fact, the door was locked, which seemed a little strange. Nonetheless, right there on the door was a schedule, which did indeed say, "Introduction to Mass Media Telecourse Orientation" in the 5-6 block.

Now what was I going to do? A telecourse depends completely on that orientation session, which is the only time the students see the instructor. Well, I would just have to call each of my students and straighten things out.

When I got home, I started calling.

"Hello. This is Diana Coogle, your instructor for the Media Waves telecourse. Did you go to orientation tonight?" (This opening would help me save face, in case the student had made a mistake.)

He thought he had. "Oh, my god!" he said, "I thought it was tomorrow night!"

I started to say sympathetically that such mistakes were easy to make and to explain my own, but I thought the better of it and merely told him where I would leave the orientation material. Then I hung up and called the next student.

"Oh, no!" she said. "I thought orientation was tomorrow night."

Uh-oh. Something sounded suspicious. As soon as I hung up, I looked at my calendar again. And, yes, right there in the square for Monday: orientation at Grants Pass, 6-7, and, two squares down the page, on Wednesday, orientation at Medford, 5-6. Not Tuesday. Wednesday. I wasn't an hour late for class; I was twenty-three hours early.

Being wrong about the time would have been a terrible mistake except that being wrong about the date, too, nullified that error. I was greatly relieved. But the best thing is not the comfort in knowing that no one would have to know about my mistake or that it had no bad consequences but the gleeful conclusion that, in fact, two wrongs can make a right.

Headache Demons with Claws of Pain

I wake up feeling sluggish and out of true. A thickness hangs about me. "Oh, please, don't let it be a headache," I think, but I know it's a futile thought. At least I don't have to teach today. Should I take an aspirin? Sometimes it helps, but most of the time pain medication can't touch a migraine.

I corkscrew, twist, and turn my head, trying to work loose the tight muscles knotted up in my neck and shoulders like Gordion's own handiwork. With every move, my neck crackles and snaps like fire. In places it sounds like sand ground underfoot; other times, it snaps suddenly, like a string bean. Rubbing the knot in my shoulder sends a claw of pain into my arm. When I hang my chin to my chest, the claw tightens in my back.

By now the hot-poker stabs of pain in my head, as though attached by wire to my stomach, provoke nausea. Trying to suppress my moans, or, sometimes, eschewing valor to allow myself the indulgence of vocal reaction, I lie down on the couch. My hand grips my neck, pressing on those fiery knots as though to push them out of my body.

When the pain surfaces behind my eye, I know the game is up: the headache demons will have this day. I get up from the couch, crying out from the sudden throbs of pain, and bank and close down the fire. Half-blind, I turn off the telephone, then stumble to the sink, where I grab a large pot (in case of vomit) and a cup of water. Then I go to bed.

The pain is digging in like devils with their picks, now behind the eye, now at the temple, always with sledgehammers at the base of the head. But now that I am in bed, I try to calm my thrashing and moaning and concentrate on the pain. Sometimes I envision sending it on the wind of my breath out of my skull to evaporate in the fresh air, but usually I simply enter the pain. I no longer fight it or desire it not to be or resent it for depriving me of a day of my life. I let it be. And then, at last, I find a semblance of peace. I am not free of pain, but I am calm. I am not relaxed, but I sleep. I wake up with spasms of nausea, but because I can't bear the lightning flashes of headache pain that vomiting induces, I am determined not to throw up. I am thrashing and moaning again, so with an effort I calm myself and regain control. I think of the time I had to conduct an awards ceremony with a migraine headache, and I think of my son, who was a guest performer with the Seattle Men's Choir once with such a headache and had to blow a conch as part of his performance. I am grateful for my bed, for my cat who lies by my side in sympathy. I am grateful for not having to be anywhere or do anything. I am in control again, and I sleep again. The next time I wake up, dusk is falling. The nausea has passed, though not the pain. I sleep again and wake up from time to time throughout the night to feel the pain pressing behind my eyes, stabbing my temples, balling up at the top of my neck. I sleep, I wake, I sleep.

Just at dawn I wake up and gingerly move my head. No claws of pain, no sledgehammers or picks. I sit up - no pain ensues. I am free again.

Gatherer of Bones

Clarissa Pinkola Estés, author of *Women Who Run with the Wolves*, tells the story of La Loba, the old woman who gathers bones and breathes life into them again. Such women, Estés tells us, are "often hairy, always fat, and ...wish to evade most company."

I am a gatherer of bones, and I live by myself in a remote cabin, but I assure you I am neither hairy nor fat, and though "old" is a relative concept, I am not old, either.

But I do gather bones. I bring them home from walks and then strew them carelessly about the yard. Bones lie under the rhododendron or on top of a stone retaining wall; they peek between flowers, lean against a tree root, lie on a doorstep. Some I keep indoors, on bookcases or window sills, but indoors or out, my bones just lie about, the way bones lie, carelessly, indifferently.

"You want psychoanalytic advice?" Estés says. "Go gather bones." That's metaphorical, of course, so I'm not sure why I want these bones around me. Maybe because bones are, as Estés tells us, the archetypal symbol of the indestructible soul-spirit. Maybe because they remind me of the life-death cycle. Or maybe because singing over the bones brings the wild creature alive again. Whatever the reason, I like my bones.

My daughter-in-law knows I'm a gatherer of bones (and she does not think of me as hairy and fat, either), so when she needed bones for a dance performance - to carry in a basket and scatter for a divination - she asked if she could borrow some of mine. I said, "Yes, of course, I would be delighted to have my bones used in that way."

Gathering them from their resting places, I considered my collection. Some bones people had given me: the bone of a stray dog tied onto a long leather thong; the claws of a rat made into earrings, the tiny toenails carefully painted red. There was a little bird skull, part of a set of tiny bones I once had. I had given the rest to a young friend at her coming-of-age ritual six years ago. Now her own bones, burned to ashes, nourish a willow tree in a garden. I had a number of antlers, one set still connected to its skull; a raptor's claw; one entire

123

backbone with attached vertebrae; and teeth: mammals' teeth, sharks' teeth, and two jawbones with teeth. I had wide, flat bones; winged bones; knobby little bones; pointed wedges of bones; bones with curved ends like musical clef signs or Doric columns. I don't know the names of my bones, but this anonymity of nomenclature increases their anonymity of being. Whatever kind of animal they held up in life, now they are all just bones.

I selected the best bones for divination and took them to the sink to wash off the mud, dirt, and other debris. I soaked them in bleach to whiten them, then laid the clean bones on top of the warm wood stove to dry.

In the midst of this project, the phone rang. "Hello, Diana. What are you doing?" the caller asked. And so I told him: "I'm washing my bones."

The Artist Is Humbled before the Banker

I think of bankers as careful, exact people who can keep lines of figures firmly in place. With scatterbrained artists and absent-minded professors, figures behave like rowdy schoolchildren with a substitute teacher. That's how my checkbook got in such a mess. Finally the figures were rampaging so out of control my only recourse was to ask a banker for help. Dread of admitting my incompetence fed my procrastination, which only gave the figures more chance for mischief, but one day at last I marched right into the bank and confessed I needed help balancing my checkbook.

"That's an expensive service," the teller, Wendy, said. "How far are you off?"

I wasn't sure, I said, squirming a little.

She looked surprised and asked for my last balance date.

I wasn't sure of that, either, I said. At that she looked so incredulous she bordered on laughter, but, taking pity on my helplessness, she went to talk to a banker of higher rank, who returned with her to the window. This was Anne, whom I had known at this bank for years.

"I understand you're having trouble balancing your checkbook," Anne said pleasantly, though her voice betrayed a twinkle of laughter. "How much are you off?"

I was forced to say again that I didn't know, and Anne looked so shocked at this slipshod way of handling one's affairs that I hastened to explain I was pretty sure there was enough money in my account to cover my checks.

A geyser of laughter seemed to be boiling just beneath Anne's banker's composure. "When did you last make a balance?" she asked, neatly, politely.

Having expected, now, this kind of exactitude, I had taken a frantic look at my checkbook while Wendy was talking to Anne at the desk and found the last balance date. Proud of being so precise, I said, "It was last August."

"August!" Anne chortled. The geyser erupted; both bankers spouted giggles.

Sweating with embarrassment, I snatched at a flotsam of honor in this wreckage of face. "It was my trip to Sweden," I said desperately. "Negotiating things in kronor - the exchange rate ...." It sounded pretty lame.

"Oh, Diana, Diana, Diana," Anne said, casting me into the camp of the hopeless.

"It would take us hours to -" Anne choked, then managed to say, "balance this checkbook" instead of "clean up this mess" and then suggested I just begin all over again.

"I tried that," I said miserably.

"I'll get you started," Anne said kindly. "Come over here with me." As I watched her work - quickly, efficiently, effectively - I told her I would rather be grading papers than doing what she was doing now.

"Do you teach your students how to balance checkbooks?" she asked sternly.

"Well, no," I said and started to explain that I taught writing in college, but she said, "Who teaches them this?" I got the point. All I could do was swallow crow pie.

"I'm just so embarrassed," I admitted at last, and she looked up, smiled, and said, "We just have to tease you a little." As she handed me my checkbook, she pointed to the new balance. "Start from here," she said, "and if you have any trouble next month, come and ask for help."

Humbly taking my checkbook, I walked out of the bank, enormously appreciative of the competence and detailed exactness of the banker, to which qualities I now added compassion, kindness, and an irrepressible laughter at the bumbling mathematics of the writer.

Trying My Best To Do without the Car

My plans were to visit Ela in Seattle for the Thanksgiving holiday, but two days before Thanksgiving the weather was stormy and rainy, and I loathed the idea of sitting in the car, driving, struggling through bad weather, all the way to Seattle. Maybe I could take the bus. But what about the fax machine I was giving Ela? How could I take it on the bus? If I drove I could take not only the fax machine but my guitar, too, and then I wouldn't miss four days of practice.

But driving would completely waste two days. If I took the bus, I could go at night, or if I traveled during the day, I could sew or read. But the Greyhound left Medford at 1:00 a.m., a most inconvenient time. My friend Tom said he would take me to the station even at that atrocious hour. That would enable me to have Thanksgiving dinner with friends in Grants Pass, too.

But the Greyhound bus, round trip, would cost $110. Driving would cost $40. Besides, the bus was bound to be crowded and unpleasant. So, what if I took the Greyhound to Seattle on Thursday and went back to Medford on Sunday on the Green Tortoise, that alternative bus line on the West coast that is more comfortable, easier, and cheaper than the Greyhound? The Tortoise leaves Seattle at a decent hour - 8:00 a.m. - but it arrives in Medford around midnight. Tom said he would meet me at the bus, but I was afraid two midnights at the bus station might put a pretty big strain on a valuable friendship.

Anyway, there was still the problem of taking the fax machine and the guitar.

Maybe I could take the train - fun, easy, smooth; I would arrive in Seattle refreshed and happy. I would have to drive to Eugene to get the train, but that wasn't nearly as bad as driving all the way to Seattle. And the train was a great deal cheaper than the Greyhound/Tortoise combination - a bargain $64 as opposed to the $88 for the bus combination. But the train south from Seattle was sold out for Sunday; I would have to take the Monday train home. I had wanted to come home on Sunday, but I could come home on Monday, instead. There was still the problem with the fax machine, but maybe it would fit in a duffel bag well buffered with blankets. At least it wouldn't have to ride in a baggage compartment. It might be a little awkward with the fax machine, my guitar, and my luggage, but I thought I could manage.

But the train arrived in Eugene at 11:00 p.m., which meant I would be driving home in the middle of the night, which is not a good idea for me because I tend to fall asleep in the middle of the night no matter what I'm doing. If I had to pay for a motel room, I might as well take the bus. Besides, I had to be in Grants Pass by 8:00 a.m. Tuesday to teach class.

Anyway, if I drove, I could also take my laundry to do at Ela's house. That would be helpful.

But I didn't want to drive to Seattle in all this rain and storm.

Maybe I could spent the night in Portland on Thursday and drive to Seattle on Friday. My friends in Portland, who would be out of town for Thanksgiving, said I was welcome to spend the night at their house; but if I did that, I would miss Thanksgiving dinner, and that seemed a little dismal.

Maybe I should take the bus. I could go into town on Wednesday to have an extra car key made that I could leave with Tom so I could drive myself to the bus station at 1:00 a.m., and he could pick up my car later. I could do my laundry in town on Wednesday, too.

But then Tom would have to involve a third person to drive him to the station to pick up my car, and, anyway, if I did that, I would be wasting another day in town doing laundry and having keys made, so I might as well drive and do my laundry at Ela's.

Maybe I should take the earlier train home from Seattle to Eugene; it arrives at 5:00 a.m. But it is more expensive, and, anyway, if I took the train, there would be no one to meet me at the station in Seattle, since Ela wasn't flying in from the East Coast until Friday at noon, and I would have to take a taxi - with my fax machine and guitar, though not, if this were the case, with my laundry - and that would be another $10 or more, and if it were going to cost that much, why didn't I just take the Tortoise and go all the way to Medford?

Because the Tortoise gets in at midnight.

Why didn't I just drive?

Because it would waste two days.

If I went into town on Wednesday I could get a book on tape from the library, so I could read while I drove, so to speak.

But that would waste a day, too.

If I drove, I could pick Ela up at the airport, which would be helpful to him. Why didn't I just drive?

Because I didn't want to drive to Seattle - and because I didn't want to miss Thanksgiving dinner with friends.

But driving was the cheapest alternative. It bothered no one to pick me up from or take me to bus or train stations; I could take my time as I saw fit. I could take the fax machine to Ela, do my laundry at his house, and not miss any practice on my guitar.

And so, in the end, I drove. Everything worked out fine, but dadgum, it's hard to give up our automobiles, even when we want to.

Safety Rules for the Driver

Rule #1. Always, always, always wear your seat belt.

Rule #2. Keep the seat belt release free of obstruction at all times.

Rule #3. Never, never, never drive when you're tired.

To my everlasting gratitude, I observe Rule #1. I have now learned #2, and never again will I disobey #3.

I spent the morning of October 15 with the wrecker getting my car out of the ditch where I had found myself when I woke up on my way home from Ashland around midnight the night before. I had waked in a massive cloud of dust and smell of hot metal as the car lurched to a stop at a dangerously precarious angle. The only thing that kept it from flipping over was the depth of the ditch, and I didn't even know the worst of it till the tow truck driver pointed out that the dents on the right side of the car couldn't possibly have been made by the side of the ditch. What actually happened was that I ran off the road on the right side and hit a tree that knocked me into a 360-degree spin and flung me in the ditch on the opposite side of the road. I'm glad I slept through it. I would have been terrified. As it was, I was scared to death just thinking I had slid into the ditch, and I fumbled and fumbled to get to the seat belt release and tried to find the emergency flicker lights and couldn't and took the keys out of the ignition but immediately dropped them because my hands were shaking so bad because I wanted *out of the car*. Because of the angle of the cliff above the ditch, I couldn't open the door far enough to get out, so I unrolled the window, climbed out, and stood on the ground, my body pressed into the car by the slant, like standing in the House of Mystery at the Oregon Vortex.

So, okay. Both legs worked, my head was in one piece, and there was no blood. A flashlight was somewhere in the jumble of things thrown to the floor of the car. I pushed on the car a few times to test its stability. It didn't budge, so I crawled back through the window into the car, found the flashlight, crawled out, opened the door far enough to reach the window handle, rolled up the window, closed the door, and started walking. The last mile marker I remembered had been three, but the first house number I came to said 8108. I was only two miles from home! (But had I slept for five miles?!) I had a flashlight; the rain was no more than a light mist; I had on a pair of good walking shoes, and best of all, both legs worked and there was no injury anywhere. Always, always, always wear your seat belt. That's Rule #1. And I will never, never, never ignore Rule #3 again.

Rule #4. Get religious so you'll have an object on which to place the immense gratitude you feel.

The cougar-protection spells I was saying all the way home probably weren't necessary. With all that adrenaline I could have fought off any cougar. I strode into the house well past midnight and was startled to find an owl, a little saw-whet, sitting on my writing desk. All the doors and windows were closed, so those who must have practical explanations for things can surmise that one of the cats must have caught it but lost it again in the house. I am content with the mystical explanation. Unlike most birds that get trapped in the house, it was not panicked and did not fly from me when I approached but only stared at me out of its wide, wise, yellow eye. I opened the window over the desk and touched the owl gently on the back of the head, suggesting immediate flight before the cats returned. The owl lifted its wings and flew through the window, over the honeysuckle and into the night.

The owl is my totem, from now till forevermore.

Through the Looking-glass

A small, round, four-inch mirror set in a vermilion-wood frame is more than a keepsake from a friend; it was for a long time the only mirror in the house. When Ela and I were here together (he lived every other year with his dad, just over the mountain), it didn't matter, as we served as each other's mirror:

"Do these socks go with these pants?"

"Is my hair sticking up in the back?"

"Is my hem straight? Do these colors match? Can I wear this with this and that with that?"

When Ela wasn't here, I had to make do with disjointed images mentally put together from what the little round mirror told me. Setting it on the window sill and stepping back as far as myopia and a small house would allow, now coming closer, now stepping farther away, bending down, rising up, I would try to make sense of what I could see a little at a time - the face, the waist, part of the legs, the feet: it was difficult to get a whole picture. Is it any wonder that after ten years of fragmented reflections I hankered for a real, full-length mirror? Finding one at a bargain price one day at Bi-mart, I bought it and carried it home.

I knew already where I wanted to hang it: to the side of the windows at the entrance to the pantry, but when I set it there, it seemed too bold, too brash, too blatantly apparent, too vocal. It shouted its presence - and mine as well. Did I really want to see myself every time I walked into the pantry or climbed down the ladder of my loft? How did people dare have mirrors all over their bedrooms and living rooms and bathrooms? Does a mirror increase a woman's vanity, or does it force stark reality onto her unwillingness to watch the aging process? Would it create honest self-appraisal or envy and jealousy? "Mirror, mirror, on the wall, who's the fairest of them all? And if there's one more fair than I, I'll tear her hair and scratch her eye."

I hung the mirror behind a curtain in the other room, figuring I could lift the curtain whenever I wanted to see if my colors matched. But that was troublesome. The mirror wasn't serving as a voice. It had to be out of the closet. Bravely one day I took it out from behind the curtain and hung it by the pantry door. I was ready at last to bear the murmur of the mirror at all times. "Mirror, mirror, on the wall, am I too fat, too short, too tall?" Go ahead and speak, I challenged it grimly, and turned my back on it.

Immersed in my work, I forgot about the mirror. When I crossed the room a few hours later, my own moving image startled me. I could be so foolish only once, though, and the mirror no longer frightens me. My outfits match, and I can tell whether my skirt needs ironing or my hair needs combing. I climb out of the tub and see the Degas image of myself.

Though it is an ordinary mirror, not a magic mirror, and though it tells the truth as any mirror does, my mirror is not, as I had suspected it would not be, a passive voice, a mere reflection. It speaks boldly and often, but it speaks with a surprisingly gentle tone. "Mirror, mirror, on the wall, who am I, who am I at all?" and the mirror, reflecting on the question, says, "This is you" and "This is you" and "This, too, is you," and the reflection reflects on me and I on the reflection till the outside shows the inside and either side could be through the looking-glass.

Becoming a Progressive Housewife

"Undoubtedly your wife wishes to be considered among the progressive members of her community," says a 1908 ad for the "ideal vacuum cleaner," which, according to the picture, has a motor bigger than my generator. The "progressive wife" (she cleans house; he does the buying) holds gracefully onto the long handle, but this is not an action photo, since the wife is not bent over with the effort, arms, back, and legs straining, ears dumb with the roar. It's a wonder she didn't prefer the broom.

I did for years. The quiet, slow broom invites contemplation and careful attention to the task. Too much exertion sends the dirt scattering, so one gently urges the dirt into place, then stoops to its level to brush it carefully into the dust pan. The broom is Zen. I am Zen. Nonetheless, when I noticed one day that some small vacuum cleaners could be run on only seven amps, a light bulb went on in my head. I could run one of those off my generator. I could carry it up the ladder to clean my loft. I could cut my cleaning time in half. I could - and did - buy a Hoover vacuum cleaner.

Dirty old London saw the first vacuum cleaner in 1901, patented by Hubert Cecil Booth. A lighter, more widely distributed model was patented in 1908 by the Americans Murray Spengler and his cousin, Hoover. To my grandmother, Hoover was another word for a vacuum cleaner.

"Its ability to suck up dust brought remarkable results, greatly simplified cleaning, and even improved sanitary conditions," says the encyclopedia about the 1908 model, but it could just as well be describing my home improvement. The thing just sucks up dirt. I point the crevice tool - thwoop! Spider webs disappear into a vacuum. I sweep over the floor. Thwoop! Grains of dirt, fir needles, and tiny shreds of bark are gone. Thwoop! The flies on the window tumble down the tube like Alice down the rabbit hole. Thwoop! Thwoop! Thwoop! Schlurp. Schlupp. All down the vacuum. Nature abhors a vacuum. I love a vacuum. Suck it all down, little Hoover! You little puppy-dog-like domesticated servant trundling obediently at the end of the flexible hose, suck it all down. Dig that hungry emptiness deep into the rug, and eat that dirt.

But my mechanized beast does have a loud growl. What happened to Zen?

Defenestrated without a tear. Zen isn't the only philosophy worth living. If the vacuum cleaner's noise closes out all other sound, into that vacuum might rush, perhaps, great thoughts. Could the vacuum cleaner lead to profound philosophies, the resolution of problems, ideas of profundity? If the broom is Zen, is the vacuum cleaner Shakespearean, Einsteinian, DaVincian?

This machine, called by the name of absence, owes its presence to vacancy. Ponder, then, that it is connected to "*va*cation," which is what we take when we want to get away from vacuuming, and to "*va*in," which is how we might be with our perfectly clean house, and to "de*va*state," which is the perception of the spiders and flies, and to "*va*nish," which is what is happening to the dust, and to "a*v*oid," which is what we used to do about housecleaning until we bought a vacuum cleaner.

Refrigeration

"Do you think these apricots will keep till Friday?" I asked the clerk at Farmer's Market. "I have to make a pie on Friday."

"Just put them in the refrigerator, and they'll be fine," he said breezily, making the same assumption most people make – that I have a refrigerator.

Well, I do have a refrigerator; I just don't have refrigeration. A mini-sized, propane-powered, RV-style refrigerator has sat on my back porch for years. It worked for a while, and then it didn't, and now it's a refrigerator only in the sense that I put things on top of it in the winter to keep them cold.

Last week a friend stopped by with a large tool box and a great desire to get my refrigerator working. I warned him that I thought the project was hopeless, but he looked up at me from where he was lying on the back deck with his arms under the refrigerator and said, "It doesn't have any moving parts. It can't break down."

Before he left, Jeff proved his workmanship stronger than my skepticism. The refrigerator was working. I stood there a minute trying to think of what I had that I should refrigerate, and I remembered a cut-up grapefruit I had put aside, hoping vaguely it would still be good tomorrow, so I refrigerated it.

As I was preparing dinner, I saw a bunch of carrots in the pantry, floating in a plastic bag blown up like a balloon and tied tightly. It occurred to me that a refrigerator was a better method of keeping vegetables fresh, so I put my carrots in (not on) my refrigerator. Next I went to the cupboard for the loaf of sprouted seed bread I had opened yesterday, which, today, gave off a most unsavory sourness. "Refrigerate after opening," the label said. That spoiled bread and the greasy, soft butter next to it suggested that having refrigeration could change my whole dietary life style. I could eat "refrigerate after opening" products like sprouted seed bread and jam and tofu over a period of time instead of all at once. I could buy apricots and keep them for days. I could use fresh milk instead of powdered, and I could buy eggs by the dozen and yogurt by the quart at any time of the year. My summer pie crusts would now be as good as my winter pie crusts with "chill for two hours" always available, and I could make custard pies all year long. I would no longer have to eat a whole bunch of kale one night and call it dinner; I could have a variety of things for dinner and refrigerate the left-overs. I could have mayonnaise if I wanted it and cheese and sour cream. I could even have ice cream! I began to worry about getting fat now that I have refrigeration.

But not to worry, and not to get too excited or rush down to the store and buy that Ben and Jerry's. The next morning when I got up, the refrigerator had quit working. Skepticism won the day, after all. I took the slimy butter, the bowl of grapefruit, and the bag of carrots out of the refrigerator and set them on top of it. I know how to live without refrigeration. I've done it for years.

A Woman's Mid-life Crisis

I saw a friend recently at a party, a man I hadn't seen for a long time. We had a long, wonderful conversation, but there was something he said that I didn't have a chance to answer at the time. At one point in our conversation, just before he was called to the other room to play music, he said, "Do women have a mid-life crisis, too?"

Do women have a mid-life crisis, Walter? Just listen to what Nature tells a woman in her fifties:

"So you're not going to produce any more children?" Nature says with a sneer. "Well, you're not worth much, then. Procreation is everything. All those things I gave you to make the opposite sex lust after you? Forget them. You don't need them any more. That nice high lift to the breasts? Let them fall. That blush on the cheek, that taut skin, those full red lips, that waspish figure? Forget it all. Who cares now? Let the face wash out and the lip line recede. Let the waist thicken and the thighs sag and the shapely hips go slack. Let the flesh hang, let the skin wrinkle. Who cares now? Go ahead; be an elephant. Those lovely tresses with their sheen and shine? No need; let them thin out, lose their color, turn brittle. Those healthy, strong bones? Let 'em break if they're going to. You're of no more use to me, so I toss you aside. Rot, go to ruin, try to recover my gifts with lipstick, bras, and hair dye. Do what you like. You don't interest me any more."

But indifference isn't enough for Nature; she also turns vindictive. "The curse" is not a woman's period but the cessation of that period. Listen to Nature's curse, Walter:

"May you wake up night after night in a sudden heat, sweating furiously," Nature begins. "May you throw off the covers and splay out your limbs, only to shiver minutes later with a sudden chill. May your nights alternate between being dream-wrenched and sweat-drenched, and may your days be plagued with internal combustion. At any time during the day may these waves of fire wash over your body; may your face blush red and shine with sweat and then grow mottled as a molting chicken, and may these things occur three or four times an hour in the most embarrassing of all social and professional circumstances when you can't walk outside to cooler air or fan yourself frantically or gulp cold water. May there also be a social stigma on this condition of your age so you can't say, 'Whew! Hot flash!' but can only pretend like nothing is happening or that no one will notice. May the heat be so intense that everyone notices and may it come on you desperately hot during a romantic dinner or while teaching class or playing tennis or having a guitar lesson. And on top of all that, may you one minute revel in euphoria and the next want to bite off the heads of your friends and snap at your cats, and may you be forced to admit that if the bitchiness is unreasonable, so is the euphoria. And so may it be for you on the threshold between youth and old age."

Do women have a mid-life crisis, *too*? Tell me, Walter, what on earth is a man's mid-life crisis like?

Playing Mussorgsky's "The Castle"

Don't strum the strings like a dadgum folk singer. And for God's sake don't pluck the strings; you're not playing a banjo. For each note, push the string straight down towards the wood of the guitar. Let each slide around the sandpaper-polished, rounded fingernail, clockwise, like a brook sliding around a yew trunk. Let each stroke beckon to the sound that swirls unheard inside the hollow of the guitar. To pluck a string is to pick up that sound and throw it at the world; to stroke a string downward is to coax that sound to give itself to the world.

And don't fling your left-hand fingers around the fretboard like a frenzied Rastafarian. This isn't a race to see whether the right-hand fingers or the left-hand fingers will get to the note first. Relax. You have plenty of time to reform the fingers into the shape of the next chord. Let them float from string to string, not zip to each position, then hesitate uncertainly ("Is the middle finger poised for C? Is the little finger stretched far enough for A?"), and then crash land just in time for the beat. No, let them float gently and beautifully; let them alight on the strings as gracefully as a great blue heron floating on her large wings and landing in the stream with only a tiny ripple of water, already balanced and steady on her long, thin legs.

Please! Don't choke the emanating sound, cutting it off at the throat by chopping the left hand into the chord ahead of the beat. Those fingers must land at the same time as the right-hand fingers stroke. The actions are simultaneous. That's what the beat is for: to see that all fingers work exactly, precisely together. It is not that the landing of the heron caused the water to ripple, but as though water and heron met each other, mutually inducing the beautiful, shimmering movement. The fingers of the left hand float between notes; they land at the precise moment the right-hand fingers stroke the same strings. And then the sound is released, freed, given to the world not in single units choked off and restarted, but in long, glorious successions of sound. This is legato; the heron rises and soars and lands again; the stream flows and swirls and ripples, and music pours out of Mussorgsky's "Castle."

140

Country Fair Pixie

Everyone agrees there is nothing quite like the Oregon Country Fair. Set once a year in July in the countryside along the banks of the Long Tom River, the Oregon Country Fair becomes a wacky little town with arts and crafts booths and food booths lining dirt paths strewn with straw to keep the dust down. Fair-goers crowd the gate, eager to be entertained by jugglers, musicians, stilt walkers, giant puppets, bubble blowers, and theater performers; eager to gawk at the outrageous costumes and painted faces - and bare, painted breasts - in the crowd; eager to buy their pots and paintings, hand-made clothes and hats, glassware and ironware.

For those of us who have booths at the fair, though, these are just the tourists; we are the townspeople, and we know that the real magic of the Country Fair happens after the tourists have gone home and before they come again the next day. We wander through the fair at night down the dark lanes, stopping for cheesecake where lanterns indicate the booth is open and then again where the whirl of a violin and stomp of a bass suggest a bluegrass jam in flickering candlelight. Dancers with multicolored, neon flares pinned to their bodies dance invisibly in the dark, like skeletons or comic-book stick figures. Two men in enormous lamp shade costumes glow eerily on the path; a drummer and a belly dancer attract a small crowd; Princess Darjeeling at a small table lit with a candle gives six words of advice to any passerby with a problem. The lanterns of the wanderers sparkle like stars on the streets of a darkened village.

I used to share a booth at the Oregon Country Fair with friends (I made woven bags and cloth dolls); after that I became partners with another friend to run a food booth (Down South, Naturally, with vegetarian black-eyed peas, cornbread, cole slaw, and tofu cheesecake). Now I go to the fair either to help my son, who is a performer there, or as a booth assistant for other friends.

Last summer I was helping my friends Rick and Louann. Rick sells paintings and gourd kalimbas, and Louann sells woven vests, bonsai plants, and pottery flower trays. To help prepare the booth for sales Saturday morning I was arranging flowers in the flower trays, working in the early sunshine with pink miniature roses, yellow coreopsis, and purple lavender spread on the shelf beside me. While I was working, a young girl came bounding across the meadow towards the booth, stopped in front of me, and said, "You are so beautiful."

Did she mean me? She was looking straight at me and there wasn't anyone else around, anyway, except Louann, who really is beautiful but wasn't close enough for the girl to have meant her. I kept thinking, "But I'm just wearing an ordinary dress, not even exotic Country Fair attire," and I had to fight down an impulse to snatch my reading glasses from off the top of my head and hide them in my pocket.

In the wake of my blushing astonishment, the girl said, "I bet you know a lot, too."

I wasn't sure I knew anything much, but I was more comfortable with the image of knowledge than of beauty, so I was able to talk with her for a bit, and then she left. Her name was Pixie.

Pixie came back the next day. She didn't say anything at first, just took off her hat and put it on my head. She pointed to a kalimba with a purple heart top and said that that was the color of a blush. She said her mother lived in Georgia. She said she herself had just gotten out of jail, where she had been for a month. I winced and suggested that jail wasn't a very pleasant place to be; maybe she should stay away from the boy friend she said had been responsible for her being there. She looked around at Rick's paintings and said she wished she could paint, and then she spewed out a string of images she said she would paint if she could, beautiful, sweet images, like frogs with wings under a pink moon, pictures of fantasy that avoided clichés and were so imaginatively detailed I ached for her to learn to draw and paint and maybe to write, too. She was fragile and sweet and a little lost, in the way of youth. I told her she had a wonderful imagination, that she ought to paint and write. I wanted my gift to her to be inspiration, to make her want to draw, to write children's books, to stay out of jail, to find that fulfillment I knew was possible. I took off her hat and gently replaced it on her head.

She smiled at me, and then she left.

I don't know that I had given her anything, but she had given me a gift in her words and admiration. I have brought that gift home with me and tucked it away with other treasures. I'll keep it there till I need it, and then on those days when I feel old and depressed, when I think I'm fat and my hair too grey, when living alone on the mountain makes me feel like the witch in a fairy tale, I'll take it out and look at it and know that all I have to do is stand in the early morning sun arranging flowers, and a beautiful spirit of a girl, a pixie, will come running across the meadow to say to me, "You are so beautiful."